Reproduction in mammals

Book 2: Embryonic and fetal development

SECOND EDITION **_Reproduction in mammals_**

BOOK **2** **_Embryonic and fetal development_**

EDITED BY C. R. AUSTIN

Formerly Fellow of Fitzwilliam College
Emeritus Charles Darwin Professor of Animal Embryology
University of Cambridge

AND R. V. SHORT, FRS

Professor of Reproductive Biology
Monash University, Melbourne Australia

DRAWINGS BY JOHN R. FULLER

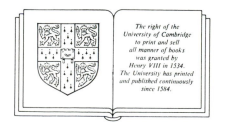

The right of the
University of Cambridge
to print and sell
all manner of books
was granted by
Henry VIII in 1534.
The University has printed
and published continuously
since 1584.

Cambridge University Press
Cambridge
New York   New Rochelle
Melbourne   Sydney

Published by the Press Syndicate of the University of Cambridge
The Pitt Building, Trumpington Street, Cambridge CB2 1RP
32 East 57th Street, New York, NY 10022, USA
10 Stamford Road, Oakleigh, Melbourne 3166, Australia

First published 1972
Reprinted 1973, 1975
Second edition 1982
Reprinted 1985, 1987

Printed in Great Britain at the University Press, Cambridge

Library of Congress catalogue card number: 81–18060

British Library Cataloguing in Publication Data
Reproduction in mammals. – 2nd edn.
Book 2: Embryonic and fetal development
1. Mammals – Reproduction
I. Austin, C.R.   II. Short, R.V.
599.01′6   QL739.2
ISBN 0 521 24786 1 hard covers
ISBN 0 521 28962 9 paperback
(First edition:
ISBN 0 521 08373 7 hard covers
ISBN 0 521 09682 0 paperback)

CONTENTS

Contributors to Book 2   vi

Preface to the Second Edition   vii

Books in the First Edition   viii

Books in the Second Edition   ix

1 The embryo   *A. McLaren*   1
Cleavage. Blastocyst formation. Experimental manipulation.
Biochemical studies. Gene action. Culture *in vitro*. Twins. Embryos in
the reproductive tract. Suggested further reading.

2 Implantation and placentation   *M. B. Renfree*   26
Implantation. Control of implantation. Placental structure and
function. Suggested further reading.

3 Sex determination and differentiation   *R. V. Short*   70
Genetic sex. Gonadal sex. Germ cell sex. Hormonal sex. Brain sex.
Phenotypic sex. Sexual selection. Behavioural sex. Legal sex. Suggested
further reading.

4 The fetus and birth   *G. C. Liggins*   114
Fetal growth. Fetal functions. Preparations for birth. Control of
parturition. Suggested further reading.

5 Pregnancy losses and birth defects   *P. A. Jacobs*   142
Causes of pregnancy wastage and birth defects. Pregnancy loss. Birth
defects. Suggested further reading.

6 Manipulation of development   *R. L. Gardner*   159
Experimental chimaeras. Manipulation of the mouse blastocyst.
Teratocarcinomas. Other uses of experimental chimeras. Investigation
of X-inactivation. Transplantation of nuclei. Suggested further
reading.

Index   181

CONTRIBUTORS TO BOOK 2

R. L. Gardner, FRS
Sir William Dunn School of Pathology
South Parks Road
Oxford OX1 3RE, UK

P. A. Jacobs
Department of Anatomy and Reproductive Biology
John A. Burns School of Medicine
1960 East-West Road
Honolulu
Hawaii 96822, USA

G. C. Liggins, FRS
Postgraduate School of Obstetrics and Gynaecology
National Women's Hospital
Claude Road
Auckland 3, New Zealand

A. McLaren, FRS
MRC Mammalian Development Unit
Wolfson House
4 Stephenson Way
London NW1 2HE, UK

M. B. Renfree,
Department of Anatomy,
Monash University,
Clayton,
Victoria 3168, Australia

R. V. Short, FRS
Department of Physiology,
Monash University,
Clayton,
Victoria 3168, Australia

PREFACE TO THE SECOND EDITION

In this, our Second Edition of *Reproduction in Mammals*, we are responding to numerous requests for a more up-to-date and rather more detailed treatment of the subject. The First Edition was accorded an excellent reception, but Books 1 to 5 were written ten years ago and inevitably there have been advances on many fronts since then. As before, the manner of presentation is intended to make the subject matter interesting to read and readily comprehensible to undergraduates in the biological sciences, and yet with sufficient depth to provide a valued source of information to graduates engaged in both teaching and research. Our authors have been selected from among the best known in their respective fields.

Book 2 deals with development from the first cleavage of the fertilized egg to the birth of the new individual. Information is provided on all the important events of normal development as well as on the anomalous processes that are known to occur, and on the experimental approach to a better understanding of both the normal and the abnormal through the use of manipulative techniques.

From the Preface to the First Edition

Reproduction in Mammals is intended to meet the needs of undergraduates reading Zoology, Biology, Physiology, Medicine, Veterinary Science and Agriculture, and as a source of information for advanced students and research workers. It is published as a series of eight small textbooks dealing with all major aspects of mammalian reproduction. Each of the component books is designed to cover independently fairly distinct subdivisions of the subject, so that readers can select texts relevant to their particular interests and needs, if reluctant to purchase the whole work. The contents lists of all the books are set out on the next page.

BOOKS IN THE FIRST EDITION

Book I. Germ cells and fertilization
Primordial germ cells *T. G. Baker*
Oogenesis and ovulation *T. G. Baker*
Spermatogenesis and the spermatozoa *V. Monesi*
Cycles and seasons *R. M. F. S. Sadleir*
Fertilization *C. R. Austin*
Book 2. Embryonic and fetal development
The embryo *A. McLaren*
Sex determination and differentiation *R. V. Short*
The fetus and birth *G. C. Liggins*
Manipulation of development *R. L. Gardner*
Pregnancy losses and birth defects *C. R. Austin*
Book 3. Hormones in reproduction
Reproductive hormones *D. T. Baird*
The hypothalamus *B. A. Cross*
Role of hormones in sex cycles *R. V. Short*
Role of hormones in pregnancy *R. B. Heap*
Lactation and its hormonal control *A. T. Cowie*
Book 4. Reproductive patterns
Species differences *R. V. Short*
Behavioural patterns *J. Herbert*
Environmental effects *R. M. F. S. Sadleir*
Immunological influences *R. G. Edwards*
Aging and reproduction *C. E. Adams*
Book 5. Artificial control of reproduction
Increasing reproductive potential in farm animals *C. Polge*
Limiting human reproductive potential *D. M. Potts*
Chemical methods of male contraception *H. Jackson*
Control of human development *R. G. Edwards*
Reproduction and human society *R. V. Short*
The ethics of manipulating reproduction in man *C. R. Austin*
Book 6. The evolution of reproduction
The development of sexual reproduction *S. Ohno*
Evolution of viviparity in mammals *G. B. Sharman*
Selection for reproductive success *P. A. Jewell*
The origin of species *R. V. Short*
Specialization of gametes *C. R. Austin*
Book 7. Mechanisms of hormone action
Releasing hormones *H. M. Fraser*
Pituitary and placental hormones *J. Dorrington*
Prostaglandins *J. R. G. Challis*
The androgens *W. I. P. Mainwaring*
The oestrogens *E. V. Jensen*
Progesterone *R. B. Heap and A. P. F. Flint*
Book 8. Human sexuality
The origins of human sexuality *R. V. Short*
Human sexual behaviour *J. Bancroft*
Variant forms of human sexual behaviour *R. Green*
Patterns of sexual behaviour in contemporary society *M. Schofield*
Constraints on sexual behaviour *C. R. Austin*
A perennial morality *G. R. Dunstan*

BOOKS IN THE SECOND EDITION

Book 1. Germ cells and fertilization
Primordial germ cells and the regulation of meiosis *A. G. Byskov*
Oogenesis and ovulation *T. G. Baker*
The egg *C. R. Austin*
Spermatogenesis and spermatozoa *B. P. Setchell*
Sperm and egg transport *M. J. K. Harper*
Fertilization *J. M. Bedford*

Book 2. Embryonic and fetal development
The embryo *A. McLaren*
Implantation and placentation *M. B. Renfree*
Sex determination and differentiation *R. V. Short*
The fetus and birth *G. C. Liggins*
Pregnancy losses and birth defects *P. A. Jacobs*
Manipulation of development *R. L. Gardner*

Book 3. Hormonal control of reproduction
The hypothalamus and the anterior pituitary *F. J. Karsch*
The posterior pituitary *D. W. Lincoln*
The pineal gland *G. A. Lincoln*
The testis *D. de Kretser*
The ovary *D. T. Baird*
Oestrous and menstrual cycles *R. V. Short*
Pregnancy *R. B. Heap and A. P. F. Flint*
Lactation *A. T. Cowie*

Book 4. Reproductive fitness
Reproductive strategies *R. M. May and D. Rubenstein*
Species differences in reproductive mechanisms *R. V. Short*
Genetic control of fertility *R. B. Land*
Effects of the environment on reproduction *B. K. Follett*
Sexual behaviour *E. B. Keverne*
Immunoreproduction *N. J. Alexander*
Reproductive senescence *C. E. Adams*

Book 5. Manipulating reproduction
Increasing productivity in farm animals *K. J. Betteridge*
Today's and tomorrow's contraceptives *R. V. Short*
Contraceptive needs of the developing world *D. M. Potts*
Risks and benefits of contraception *M. P. Vessey*
Augmenting human fertility *D. T. Baird*
Our reproductive options *A. McLaren*
Barriers to population control *C. R. Austin*

ix

1
The embryo

ANNE McLAREN

The mammalian egg, though small compared with the eggs of other vertebrates, such as fish, amphibians, reptiles and birds, is one of the largest cells in the mother's body. Before fertilization, it is also one of the most inactive cells, with a metabolic rate scarcely higher than a very inert tissue such as bone. It possesses individuality, in the sense that its complement of genes constitutes a unique selection from the maternal set, but its genetic information has not yet been used and hence in all inherited characters it resembles other maternal cells.

Three to four days after fertilization of the egg, the embryo may still contain less than a hundred cells, but these cells now approximate to those of an adult organism in average size. The patterns of RNA synthesis and the capacity for synthesizing protein are also similar to those of the adult, while the metabolic rate is as high as the most metabolically active tissues of the maternal body, such as the retina. The embryo's own genes have begun to function, coding for proteins that are potentially recognizable as foreign by the mother. Differentiation has begun, giving rise to a tissue, the trophectoderm, highly specialized for interaction with the mother.

These striking changes form the main subject matter of this chapter. Since much of what we know about mammalian embryos comes from experiments in which they have been cultivated outside the mother's body, the process of cultivation *in vitro* will also be described, together with the possibilities it offers for experimental analysis and manipulation.

Cleavage

The diameter of the egg is about 0.1 mm in all eutherian mammals and marsupials – just at the limit of visibility to the unaided eye. It is surrounded by a thin non-cellular envelope, the zona pellucida (Fig. 1.1). The one-cell stage, because of its relatively large size, is distinguished by a low ratio of nuclear to cytoplasmic material. This ratio, critical for the genetic control of the cell at later stages, is restored to a value resembling that of adult cells by the process of cleavage, in which several successive cell divisions occur without any increase in total mass. Indeed, growth may be 'negative' during cleavage: the total amount of cellular material decreases by about 20 per cent in the cow, and by as much as 40 per cent in the sheep, while the protein content of the mouse embryo falls by 25 per cent during the first 3 days after fertilization.

Cleavage takes place much more slowly in mammals than in most lower vertebrates or invertebrates. Frog eggs cleave about once an hour, and goldfish eggs every 20 minutes or so; but a mouse egg takes 24 hours for its first cleavage division, and 10–12 hours for each succeeding division. Other mammalian embryos that have been examined show the same slow tempo. We do not know the reason for this difference.

Variation in rate of cleavage is common, both among embryos and among the cells (blastomeres) of a single embryo. One consequence is that the initial synchrony of the cleaving embryo soon vanishes. Embryos with 2 and 4 cells are much more often encountered than those with 3 and 5 cells; the following day, 8-cell stages predominate, but the scatter is wider; after

Fig. 1.1. Early development of the embryo in eutherians and marsupials up to the time of implantation. (Adapted from W. A. Wimsatt, *Biol. Reprod.* **12**, 1–40 (1975).)

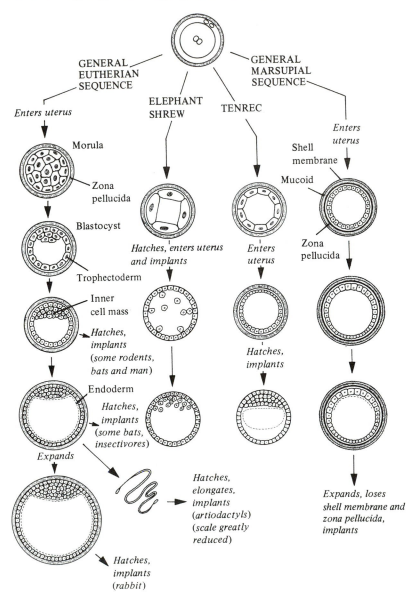

four or five successive cleavage divisions, little synchrony remains. The first cell to divide from the 2-cell stage mouse embryo has recently been shown by Chris Graham and his colleagues in Oxford to contribute a disproportionately larger number of progeny to the inner cell mass of the blastocyst, and fewer to the outer trophectoderm.

During the first few cleavage divisions, each mitosis is followed immediately by DNA synthesis in the two daughter cells; this means that the pause (G1 period) before DNA synthesis, which is characteristic of adult cells, is absent. No G1 period can be detected in mouse embryos before the 8- to 16-cell stage.

Blastocyst formation

At some stage during early cleavage (late 8-cell in the mouse), a striking and significant alteration occurs. The cells change from spherical to wedge shape and flatten against one another so as to maximize the amount of cell-to-cell contact; they then develop special links or junctional complexes. This whole process, termed compaction, gives the cells a polarity for the first time, in that each now has an outer and an inner end.

Within this solid ball of cells, the morula, a fluid-filled cavity appears. The cavity (blastocoele) enlarges rapidly, until the embryo resembles a hollow sphere, the blastocyst, with a single peripheral layer of large flattened cells, the trophectoderm or trophoblast, and a knob of smaller cells to one side of the central cavity. This knob, the so-called inner cell mass, will give rise mainly to the adult organism, while the cells of the trophectoderm form the placenta and embryonic membranes. However, it is a characteristic of all marsupial embryos that there is no inner cell mass, and the blastocyst appears as a hollow ball of cells, all similar to one another. In marsupials, we do not know what decides which cells will go to form embryo and which to form embryonic membranes.

Up to this stage, the embryos of all eutherian mammals resemble one another closely. Blastocysts, however, develop rather differently in different groups. In the mouse, and probably also in man, the blastocyst cavity begins to form when no more than 20–30 cells are present, and at the time that implantation begins the blastocyst contains no more than 100 or so cells. Marsupials are similar. In rabbits, blastocyst formation begins about three cleavage divisions later than in the mouse, while the mature blastocyst contains several thousand cells and measures 3–4 mm in diameter. In sheep and cattle, the blastocyst gradually elongates, and may attain a length of 20 cm before attachment in the second or third week of pregnancy. In pigs, the elongating process is accentuated: between the ninth and sixteenth day of pregnancy the blastocyst undergoes a 300-fold elongation, changing from a small spherical vesicle to an exceedingly long thread-like tube more than a metre in length, before attachment begins. (Fig. 1.2).

The time at which the blastocyst cavity appears is not related to the size of the embryo. If three blastomeres in a 4-cell mouse or rabbit embryo are destroyed, the survivor continues to develop; the blastocyst cavity will form at the usual time, giving rise to a blastocyst a quarter the size of a normal one. If an embryo is disaggregated into single cells, a fluid-filled vesicle often appears in individual cells at the time when the blastocoele would normally form, suggesting that the fluid originates by intracellular accumulation. The timing of fluid secretion is not related to the number of cleavage divisions that have elapsed since fertilization, since it does not change when one cell division is experimentally suppressed. It may however depend on the number of mitotic divisions that have taken place.

The most striking feature of the blastocyst is its differentiation into trophectoderm and inner cell mass. Differentiation is the central unsolved mystery of development, since from a single fertilized egg all the tissues of the adult organism develop, including such diverse types as bone, brain, spermatozoa, and pigment-forming cells. Trophectoderm cells appear relatively specialized in that they are large and flattened with numerous microvilli, forming the continuous close-knit wall of the blastocyst (Fig.

Fig. 1.2. (*a*) Two-celled pig embryo from the Fallopian tube about 18 hours after ovulation, with spermatozoa embedded in the zona pellucida. (*b*) Eight-celled pig embryo from the uterine horn about 55 hours after ovulation. The blastomeres are all similar in size and appearance. (*c*) Pig blastocyst 13 days after ovulation. The total length of the blastocyst was 157 cm, and the position of the embryonic disc is arrowed.

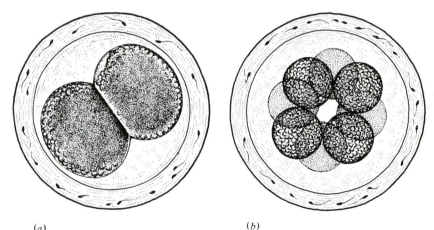

(*a*)   (*b*)

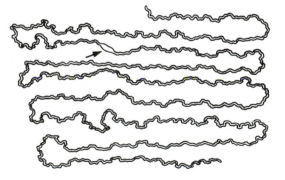

(*c*)

1.3). They include about two-thirds of all the cells of the blastocyst. Electron microscope investigations show that their membranes are closely apposed and interdigitated, linked at intervals by the 'tight junctions' characteristic of epithelial tissues. These tight junctions have been observed as early as the 8- to 16-cell stage, around the periphery of the embryo. Functionally, the trophectoderm acts first as a pump, in that it is responsible for the active transport of ions (e.g. Na^+, Cl^-) between the outside and the blastocyst, and thus indirectly for the transfer of fluid into the cavity. Later, the trophectoderm gives rise to primary trophoblast cells, which are the first to make contact with maternal tissue, ingesting the disintegrating uterine epithelium and invading the deeper layers. In contrast, the inner cell mass consists of small, rounded, rapidly dividing cells, showing minimal mutual adhesion or specialization until after implantation has occurred. Differentiation has also occurred at the biochemical level: Martin Johnson and his colleagues in Cambridge have shown that inner cell mass and trophectoderm synthesize different populations of proteins. The causal basis that underlies this first and very fundamental step in differentiation is discussed by Richard Gardner in Chapter 6.

In some species, such as the goat and pig, larger cells at one pole can already be seen in the morula: these constitute the embryonic ectoderm, while the trophoblast develops separately and does not enclose the

Fig. 1.3. A scanning electron micrograph of a mouse blastocyst. The closely compacted appearance of the cells is attributable to the tight junctions that have developed between them. × 800.

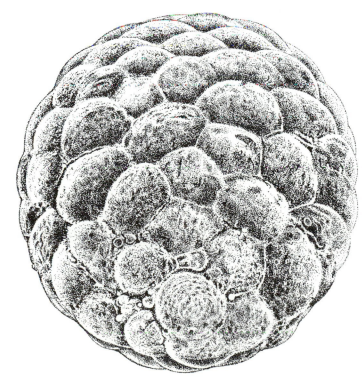

embryonic mass (Fig. 1.4). In these forms, therefore, to speak of an 'inner cell mass' is misleading.

The next tissue to differentiate is the primary endoderm, which forms from the layer of cells facing the blastocyst cavity in the inner cell mass. Endoderm cells migrate around the entire inner surface of the trophectoderm. After implantation, they perform a vital nutritive role for the early embryo, but they do not give rise to the definitive endoderm of

Fig. 1.4. Formation and development of the blastocyst in the goat. 1. Large cells representing the future embryo become evident at one pole of the morula and fluid-filled spaces make their appearance among the smaller cells. 2,3. The blastocoele enlarges, the embryonic cells form a rounded mass and the endoderm differentiates. 4,5. The embryonic 'plate' forms on the surface of the blastocyst. 6. The neural groove can be distinguished and also some mesoderm cells beneath the embryonic plate. Further progress in development is discussed in Chapter 2.

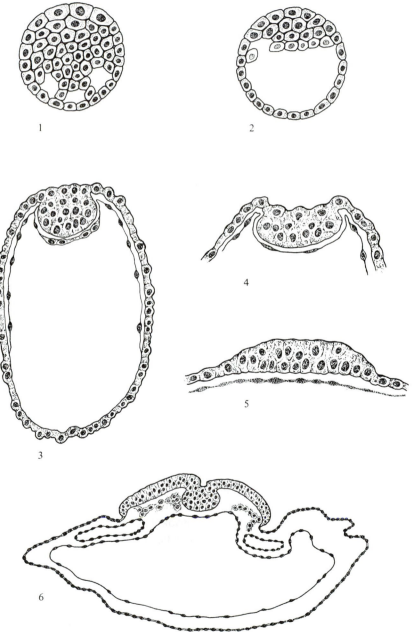

the fetus. The remainder of the inner cell mass is now termed the epiblast; at gastrulation it forms the primitive streak and gives rise to the ectoderm, mesoderm and definitive endoderm.

Experimental manipulation

Until about 1960, very little experimental embryology had been done on mammals because of the small size and inaccessibility of mammalian embryos. During the last 20 years, these obstacles have been largely overcome: not only mouse embryos, but also rabbit, sheep, cow and human embryos, can now be cultivated *in vitro* throughout the cleavage period, and subjected to a variety of experimental procedures. After cultivation *in vitro*, embryos may be transferred to the uterus of a female at an appropriate stage of pregnancy or pseudopregnancy, so that their developmental potential can be assessed.

Early claims that parthenogenetic activation of unfertilized rabbit eggs was followed by full normal development have not been confirmed. Parthenogenetic activation of the eggs of several mammalian species has certainly been achieved by a variety of methods, but development usually ceases at or before the blastocyst stage. Only in the mouse has parthenogenetic activation been followed by development up to mid-gestation. The methods used have included electric shock and treatment of eggs with alcohol or media low in calcium. Some of the embryos develop as haploids, with half the normal chromosome number, some as diploids, and some as haplo-diploid mosaics. Why is development arrested at mid-gestation? Tissue from parthenogenetic embryos shows extensive growth on transfer to sites outside the uterus, and will participate in organogenesis and survive into the adult, even giving rise to functional germ cells, if aggregated with normal embryonic cells to form a chimaera (see Chapter 6). A sperm nucleus is not required for normal development: Peter Hoppe and Karl Illmensee in the USA have removed the male pronucleus from fertilized mouse eggs by microsurgery, and after doubling the chromosomes of the egg by treatment with cytochalasin to suppress one cell division, have produced adult mice from the resulting embryos. Indeed, we know that the parthenogenetic nucleus itself can support normal development: the same two workers, in Geneva this time, have transferred a nucleus from a parthenogenetic embryo to an enucleated mouse egg, and have again produced adult mice. The cause of death of parthenogenetic embryos therefore remains a mystery. (Other aspects of parthenogenesis are discussed in Book 1, Chapter 3.)

Triploidy has been induced in mice, rats and rabbits by suppression of second polar body formation. The triploid embryos develop apparently normally to mid-gestation, and then die. Spontaneous triploidy has been detected in mouse embryos, and is a significant cause of embryonic mortality in man. (Triploid development as a consequence of disorders of fertilization is dealt with in Book 1, Chapter 6, and is discussed further

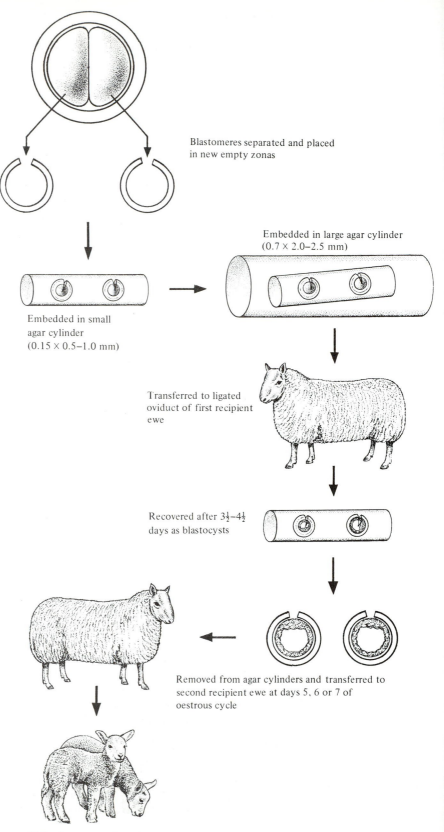

Fig. 1.5. Technique for producing identical twins, triplets or quadruplets by separating the blastomeres of early cleaving eggs, transferring each blastomere back into an empty zona pellucida, and embedding it in agar before transfer back into the oviduct of a recipient. The embryos are recovered at the blastocyst stage, removed from the agar cylinders and transferred to the uterus of a second recipient for further development. (From Willadsen, 1979 – see Suggested Further Reading). (*Opposite page*). Three sets of twin lambs produced by this method. (Photo by courtesy of S. M. Willadsen).

Blastomeres separated and placed in new empty zonas

Embedded in small agar cylinder (0.15 × 0.5–1.0 mm)

Embedded in large agar cylinder (0.7 × 2.0–2.5 mm)

Transferred to ligated oviduct of first recipient ewe

Recovered after $3\frac{1}{2}$–$4\frac{1}{2}$ days as blastocysts

Removed from agar cylinders and transferred to second recipient ewe at days 5, 6 or 7 of oestrous cycle

Birth of monozygotic twins

in Chapter 5 of this Book.) Tetraploidy is less common than triploidy in human embryos and even more lethal. Some tetraploid mouse embryos, however, produced experimentally by Mike Snow in Edinburgh by suppressing one cell division in early cleavage, have survived to birth.

The effect of destroying one or more cells of the early embryo has been investigated. When one cell of the 2-cell mouse embryo is destroyed, the surviving blastomere usually gives rise to a normal blastocyst, though with half the usual number of cells. After transfer to a foster-mother, development continues normally, culminating in a full-sized, fertile adult mouse. Regulation of size is achieved about half-way through gestation. When three cells of the 4-cell stage are destroyed, an abnormal blastocyst frequently develops, consisting only of a shell of trophectoderm, with no inner cell mass. Such a structure is incapable of further development. When seven cells of the 8-cell stage are destroyed, these abnormal trophectodermal vesicles predominate. In rabbits, on the other hand, normal young have been born after destruction of seven out of eight cells. Identical twins and even quadruplets have been produced experimentally in sheep by Steen Willadsen in Cambridge, by separating the blastomeres at the 2-cell or 4-cell stage and allowing each to develop independently. Each isolated blastomere was replaced in an empty zona pellucida and then embedded in a cylinder of agar to seal the zona and protect the embryo following transfer to a temporary recipient. The embryos were recovered again after a few days

Fig. 1.5 (*cont.*)

and the agar removed to allow transfer to a definitive recipient ewe, where pregnancy proceeded normally (Fig. 1.5).

Embryos can not only be taken to pieces, they can also be joined together. Techniques for removal of the zona pellucida and aggregation of cleavage stages (most conveniently, 8-cell embryos) to produce chimaeras were developed independently by Andrzei Tarkowski in Warsaw and by Beatrice Mintz in Philadelphia. Chimaeras are discussed by Richard Gardner in Chapter 6.

Biochemical studies

Metabolic rate, whether judged by oxygen uptake or carbon dioxide output *in vitro*, increases little during the early cleavage divisions, but rises sharply between the morula and blastocyst stages in both mouse and rabbit (Fig. 1.6). In the rabbit, this is partly due to a shift from the hexose mono-

Fig. 1.6. CO_2 output and RNA and protein synthesis in the pre-implantation mouse embryo.

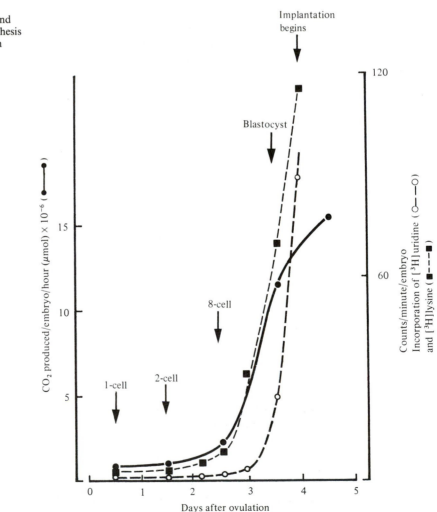

phosphate oxidation pathway to the more efficient Embden–Meyerhof pathway and the tricarboxylic acid cycle.

All mammalian embryos are probably dependent on a continuing supply of energy from the maternal environment, unlike frog or sea-urchin embryos, which can develop in pure water or a simple salt solution, respectively. The compounds that the mouse embryo is able to utilize as an energy source *in vitro* change greatly as development proceeds (Table 1.1). The fertilized 1-cell egg resembles the oocyte in requiring pyruvate or oxaloacetate; the 2-cell embryo can also utilize phosphoenolpyruvate and lactate; but it is not until the 8-cell stage onwards that glucose can be used as a source of energy. If follicle cells are present around the fertilized egg, development is possible from the very earliest stage with glucose as the sole energy source in the culture medium, since the maternal cells convert the glucose into pyruvate. The process of biochemical differentiation during cleavage may be related to an alteration in the ultra-structure of the mitochondria: 1- and 2-cell stages contain rounded mitochondria with few cristae like those of developing oocytes, while from the 8-cell stage onwards the normal adult type is present with abundant cristae.

A large increase during cleavage in the amount of glycogen in the cytoplasm has been shown in the mouse. Whether this is used as an energy source at a later stage of development is not known.

The amount of RNA in the early mouse, rat or rabbit embryo has been measured by biochemical and cytophotometric (Azure-B staining) methods. The unfertilized egg is rich in RNA, and during the first few cleavage divisions the total amount of RNA increases little or even declines. Since studies *in vitro* on the incorporation of radioactively labelled uridine into embryos indicate that some RNA synthesis is taking place at this time, breakdown of RNA must also occur. The rate of incorporation of uridine increases as development proceeds, but some has

Table 1.1. *Energy sources capable of supporting the development* in vitro *of mouse eggs and embryos at different stages of development*

	Stage of development			
Substrate	Oocyte	1-cell	2-cell	8-cell
Pyruvate	+	+	+	+
Oxaloacetate	+	+	+	+
Phosphoenolpyruvate	−	−	+	+
Lactate	−	−	+	+
Glucose	−	−	−	+

Source: From J. D. Biggers and R. M. Borland (1976) – see Suggested Further Reading.

been shown from the 2-cell stage onwards in mouse, rabbit and hamster embryos. Label is confined to the nucleus at the 2-cell stage, appearing for the first time in the cytoplasm at the 4-cell stage. Nuclear label at the 4-cell stage is concentrated over the nucleoli, suggesting that ribosomal RNA is being synthesized. For the 4-cell mouse embryo, this has been confirmed by column chromatography and sucrose gradients, which show large radioactive peaks of 28 S and 16 S material. Also at the 4-cell stage, nucleoli of adult type are first seen. It has been calculated that by the blastocyst stage 10 000 new ribosomes are made every minute in each cell of a mouse embryo, a rate of production comparable to that of adult cells in culture.

Similar techniques have been used to demonstrate the synthesis of messenger RNA and low-molecular-weight RNA, both 5 S and 4 S (transfer RNA), from the 2-cell stage onwards. The three major classes of DNA-dependent RNA polymerases have all been identified in pre-implantation mouse embryos.

Mammalian embryos differ strikingly from most other embryos (such as those of amphibians, teleost fish, echinoderms, insects) not only in requiring an external energy source, but also in the precocious onset of RNA synthesis. Both differences probably reflect the small size of the mammalian embryo and the very limited nature of its nutrient reserves. In the mouse, ribosomal, messenger and transfer RNAs are produced as early as the 2-cell stage, and by the blastocyst stage the rate of synthesis of these RNA classes is indistinguishable from that of an adult cell. The frog embryo does not achieve this degree of biochemical maturity until the gastrula stage, by which time it contains 30 000 cells. Because early RNA synthesis is necessary for normal development, mammalian embryos are very susceptible to the action of actinomycin D which inhibits the formation of RNA: a 10^{-7} M concentration in the culture medium is sufficient to reduce RNA synthesis in the mouse by 90 per cent. By contrast, amphibian and echinoderm embryos, in which RNA synthesis is so much more retarded, are highly resistant to the harmful effects of actinomycin D during cleavage and blastulation; throughout these stages they depend upon RNA elaborated in the course of oogenesis.

Protein synthesis, as judged by the incorporation of radioactively labelled amino acids, can be detected in cultured mammalian embryos from the fertilized egg onwards. Since the total amount of protein has been shown in the mouse embryo to decline during the first few cleavage divisions, breakdown of protein must occur to a greater extent than synthesis. Two-dimensional gel electrophoresis shows that some new proteins start being synthesized at fertilization or activation of the egg, but in the mouse the most striking qualitative changes occur at the 2- to 4-cell stage. The rate of protein synthesis increases dramatically from the morula to the blastocyst stage, but the types of protein being synthetized do not change so much at this time.

Little is known about the synthesis of individual proteins during the first few days of life. Spindle proteins, necessary for the formation of the mitotic apparatus, are probably produced at this time since cleavage can be prevented by treatment with a specific protein synthesis inhibitor, puromycin. The activities of some enzymes during the pre-implantation period have been determined by Ralph Brinster in Philadelphia and by Charles Epstein and his colleagues in San Francisco. Lactate dehydrogenase is present in mouse embryos in large amounts during the first 2 days of development, and thereafter declines exponentially, presumably owing to selective degradation. The sex-linked enzyme glucose-6-phosphate dehydrogenase (the gene for which is carried on the X chromosome) shows the same pattern of activity. Another X-coded enzyme, hypoxanthine phosphoribosyl transferase, shows a slow increase in activity during early cleavage, and a more rapid increase from the morula stage onwards. Other enzymes, including malate dehydrogenase, isocitrate dehydrogenase and an aldolase, show a modest increase in activity, about 30 per cent in 4 days, which is probably due to a low rate of enzyme synthesis. Hexokinase and two phosphoribosyltransferases show a large increase in activity during this period. The patterns of activity appear to be similar *in vivo* and *in vitro*, for all enzymes studied.

Gene action

An important problem concerns the stage at which the embryonic genome first becomes active. At what stage is the DNA of the embryo first transcribed? At what stage is embryo-derived messenger RNA first used in protein synthesis, as opposed to long-lived messenger RNA produced during oogenesis? At what stage are paternal genes first expressed?

The evidence, summarized above, on RNA synthesis during early cleavage of mouse embryos suggests that embryonic DNA must be functional from the 2- to 4-cell stage. The embryo-derived messenger RNA must be put to immediate use, since paternal gene expression has been detected at the 2-cell stage for at least three genetic factors, namely the genes *agouti*, *yellow* and t^{12}, and the albino locus deletion *c25H*. Paternal variants of several enzymes and antigens, further evidence of embryonic gene action, are expressed at the 6- to 8-cell stage in the mouse.

On the other hand there is also evidence that maternal messenger RNA transmitted in the egg cytoplasm is used during early cleavage. Some of the new proteins first synthesized in the 2-cell mouse embryo have been shown by Peter Braude in Cambridge to be coded for by maternal messenger RNA, present in the unfertilized egg but in a masked, inactive form. Ralph Brinster and his colleagues in Philadelphia found that globin messenger RNA injected into fertilized mouse eggs was still directing globin synthesis 24 hours later, but by the blastocyst stage no further globin synthesis could be detected, in striking contrast to the situation in the egg of the toad *Xenopus* where an injected globin message retains its activity for at least 2 weeks.

Inter-species hybrid embryos usually undergo normal cleavage, but may fail at the blastocyst stage: for example rabbit *Oryctolagus cuniculus* × hare *Lepus europaeus*, and rabbit × cottontail rabbit *Sylvilagus floridanus* (see Book 6, Chapter 4 in the First Edition of this series). In rats, Andrzrej Dyban and his colleagues in Leningrad have induced chromosome aberrations by X-irradiation of the male gametes or treatment of the zygotes with antimetabolites, and obtained embryos with varying degrees of aneuploidy, i.e. loss of one or more chromosomes. Even where the missing chromosome was one of the largest of the set, cleavage and blastocyst formation occurred normally, the embryos usually dying at about the time of implantation. A similar result has been obtained with aneuploid mouse and rabbit embryos.

Another method of producing embryos with one chromosome too few or too many has been developed by Alfred Gropp. It is based on crossing house mice *Mus musculus* with tobacco mice *Mus poschiavinus* in which the chromosomal material is differently arranged (see Fig. 1.7). Spermatogenesis in the hybrid males is characterized by a high rate of non-disjunction. The embryos that lack a chromosome (monosomic) again die at about the time of implantation, while those with an extra chromosome (trisomic) survive into the post-implantation period, but die before birth and are often grossly abnormal (Fig. 1.8).

The balance of evidence suggests that both maternal and embryonic messenger RNA are used during early cleavage, with the maternal molecules being progressively degraded, so that by the blastocyst stage the embryonic genome is in virtually complete control of development.

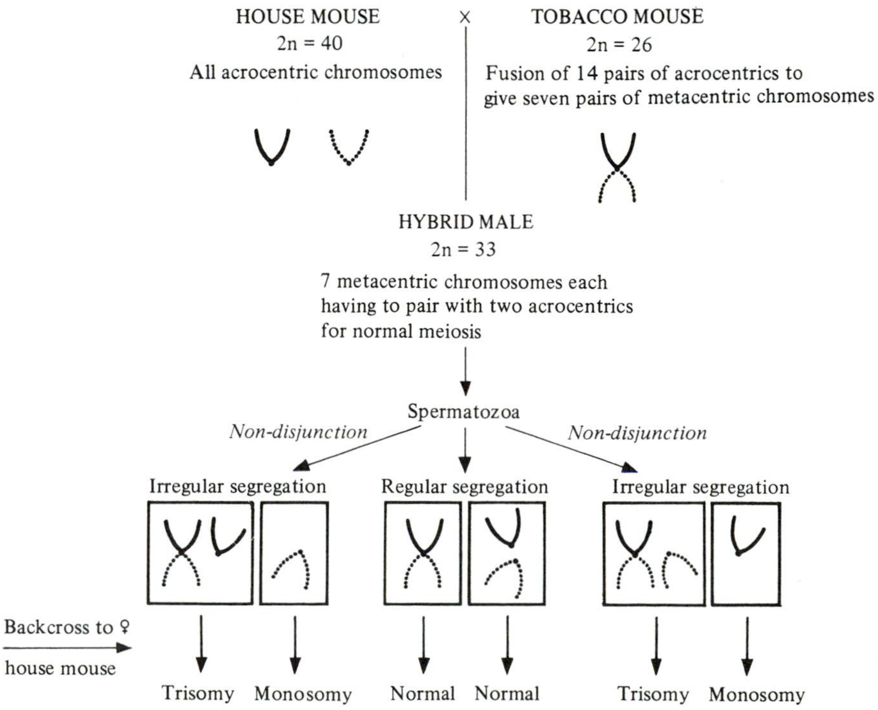

Fig. 1.7. A diagram to illustrate how it is possible to produce selective monosomies or trisomies in mice by mating hybrid males with 33 chromosomes (a cross between normal mice with 40 chromosomes and Tobacco mice with 26 chromosomes) with normal females. Meiotic non-disjunction results in the production of some chromosomally unbalanced spermatozoa containing one too many or one too few chromosomes. If these abnormal spermatozoa fertilize a normal egg, a range of monosomic and trisomic embryos will be produced.

HOUSE MOUSE
2n = 40
All acrocentric chromosomes

×

TOBACCO MOUSE
2n = 26
Fusion of 14 pairs of acrocentrics to give seven pairs of metacentric chromosomes

HYBRID MALE
2n = 33

7 metacentric chromosomes each having to pair with two acrocentrics for normal meiosis

Spermatozoa

Non-disjunction   *Non-disjunction*

Irregular segregation   Regular segregation   Irregular segregation

Backcross to ♀ house mouse

Trisomy   Monosomy   Normal   Normal   Trisomy   Monosomy

This interpretation is supported by recent studies on X-chromosome activity in early mouse embryos. Both X chromosomes are transcriptionally active during the development of the mammalian oocyte, unlike the situation in female (XX) somatic cells, where one X chromosome is always inactive. XO female mice are fertile (in contrast to XO women), but their

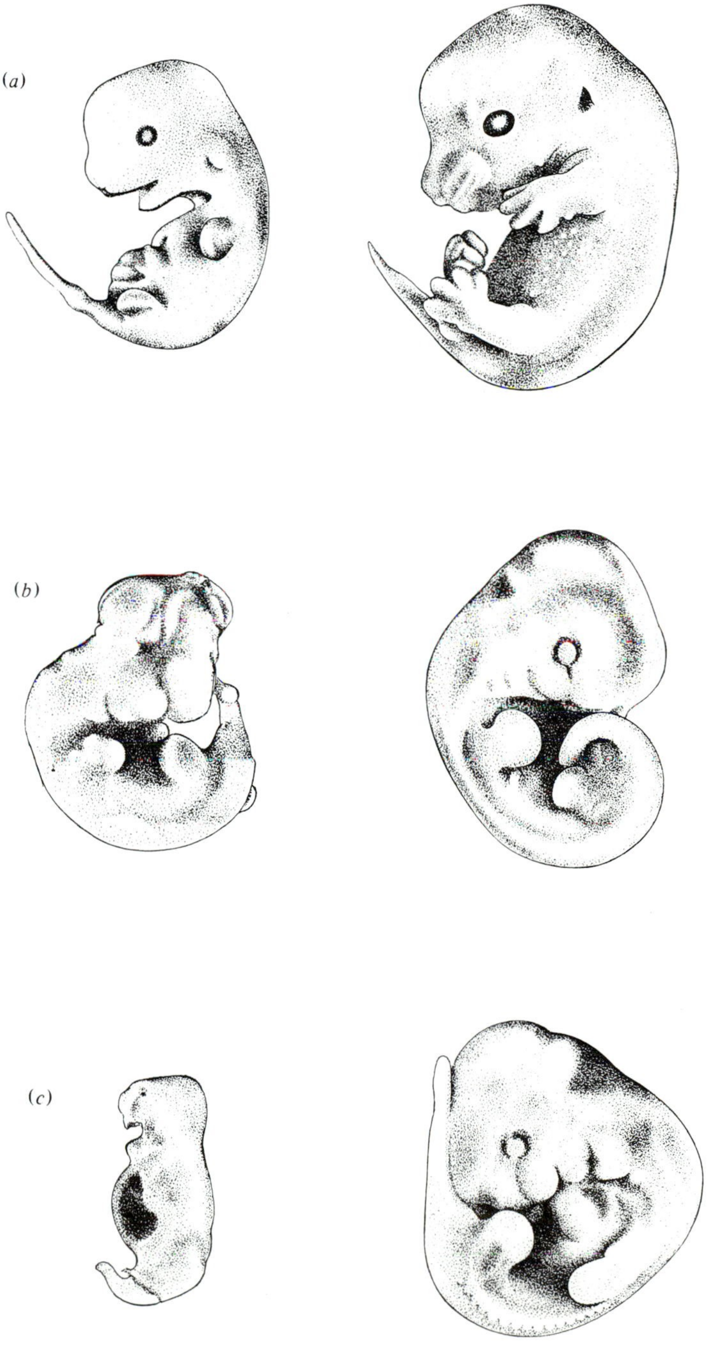

Fig. 1.8. The appearance of trisomic mouse embryos. (*a*) Trisomy of chromosome 1 and normal embryo at day 15 of gestation, showing microgenia (under-development of the lower jaw). (*b*) Trisomy of chromosome 12 and normal embryo at day 12 of gestation, showing exencephaly (extrusion of part of the brain owing to incomplete closure of skull bones). (*c*) Trisomy of chromosome 17 and normal embryo at day 12 of gestation, showing retardation and gross malformation. (From A. Gropp, D. Giers and U. Kolbus. *Cytogenet. Cell Genet.* **13**, 511–35 (1974).)

oocytes, having only a single X chromosome, show only half as much activity for enzymes coded by the X chromosome (hypoxanthine phosphoribosyl transferase, glucose-6-phosphate dehydrogenase and phosphoglucokinase) as do oocytes from normal XX females. For hypoxanthine phosphoribosyl transferase this two-fold difference in activity persists through fertilization and early cleavage, and only disappears once compaction has occurred. This suggests that enzyme activity up to the 8-cell stage is controlled by maternal gene products, and after that by the embryonic genome.

Between the 8-cell and the blastocyst stages, both X chromosomes in female mouse embryos must be functioning, since there is a two-fold difference in the levels of X-linked enzyme activities between female and male embryos. Inactivation of one X chromosome, usually that derived from the father, takes place in trophectoderm cells at the blastocyst stage, and in primary endoderm as this tissue differentiates. Marilyn Monk and Mary Harper in London have recently shown, by studying the level of hypoxanthine phosphoribosyl transferase activity, that X-chromosome inactivation in the epiblast cells of mouse embryos does not take place until just before gastrulation. In the epiblast, and hence in the mouse that it develops into, maternal and paternal X chromosomes are inactivated at random. This is in contrast to the situation in marsupials, where there is evidence for the selective inactivation of the paternal X chromosome in all tissues, not merely in the embryonic membranes.

The biochemistry of the fluid in the rabbit blastocyst cavity has been studied with radioactive tracers to determine the passage of substances into and out of the blastocyst. Before implantation the fluid is very rich in potassium and bicarbonate, which appear to be actively drawn into the blastocyst from the uterine fluid. As implantation proceeds, potassium and bicarbonate fall to the levels found in the maternal serum, while protein and glucose, previously present in small amounts only, increase up to maternal serum levels. Phosphorus and chlorides also increase greatly in concentration. Thus the blastocyst possesses a high degree of metabolic selectivity, actively controlling the rate of entry of substances from the surrounding uterine fluid.

Culture *in vitro*

Much of the information summarized above on the biochemistry of the pre-implantation embryo could only have been obtained by culturing the embryos *in vitro*, outside the maternal body in a chemically defined medium. Since such a medium contains no complex constituents, for instance serum or plasma, the nutritional requirements of the embryo can be analysed fairly easily. The first widely used chemically defined medium, devised by Wesley Whitten, consisted essentially of a bicarbonate-buffered salt solution (Krebs–Ringer) supplemented with glucose and albumin. It supported the development of mouse embryos from the 8-cell to the blastocyst stage. By subsequent transfer to foster-mothers, the author in collaboration with John Biggers showed that the developmental potential of the cultured embryos was unimpaired, in that they gave rise to normal fertile adult mice.

The period over which mouse embryos could be successfully cultured was pushed backwards in time by Ralph Brinster, who modified the medium, principally by the addition of lactate and pyruvate, so as to allow the development of 2-cell embryos which cannot utilize glucose as an energy source (see Table 1.1). Brinster also investigated the optimal range of pH and osmolarity. For several years progress was held up because development *in vitro* appeared to be blocked mysteriously at the 2-cell stage. Fertilized eggs would undergo one cleavage division and then stop unless they were returned to the reproductive tract, although late 2-cell stages would develop *in vitro* up to the blastocyst stage. The 2-cell block did not show itself if embryos were placed inside the ampullary region of the oviduct growing in organ culture in a chemically defined medium. This suggested that the ampulla provides a specialized environment in some way necessary for the early development of the mouse. Now, however, embryos of particular strains, in certain laboratories, can be cultured all the way from 1-cell to blastocyst *in vitro*, but the difficulties encountered in repeating these experiments in other laboratories show that we are still ignorant of some of the essential requirements of the culture system. The same point is still more true of other species; no one has yet been able to culture hamster eggs *in vitro* from the 1-cell stage to the blastocyst, though development can be achieved in oviducts cultured *in vitro*.

The albumin in the culture medium is probably fulfilling a physical rather than a nutritional role; it could be acting as a membrane stabilizer, since it can be replaced by a synthetic macromolecule, polyvinylpyrrolidone, permitting development of mouse embryos from the 2-cell up to the blastocyst stage, in the absence of any external source of fixed nitrogen.

Rat and rabbit embryos have also been cultured in chemically defined medium during the pre-implantation period. One of the protein fractions present in the rabbit uterus shortly before implantation (referred to both as blastokinin and uteroglobin) may be required for the expansion of the rabbit blastocyst, but recently expansion has been successfully achieved *in vitro* in a very simple medium.

Other mammalian embryos (notably pig, sheep, human) have been successfully cultured during the pre-implantation period in various media, usually containing serum, but in no species have the nutritional requirements been so precisely analysed as in the mouse.

Once the blastocyst stage has been achieved, mouse embryos have been shown by time-lapse cinematography to undergo striking cycles of contraction and expansion. A rapid contraction, thought to be mediated by microfibrils in the outer trophectoderm layer, reduces the blastocoele cavity to about one-third of its former size, and is followed by a slower expansion phase brought about by passage of fluid into the blastocoele cavity as a result of the activity of an ion pump in the wall which establishes a concentration gradient between the exterior and the cavity. Whether this cycle also takes places *in vivo*, and if so what its function is, remains unknown. If serum is added to the culture medium, the hatched mouse blastocyst attaches to the glass or plastic surface of the culture dish and the trophectoderm cells grow out as a two-dimensional sheet, undergoing differentiation into trophoblastic giant cells similar to those seen *in vivo* (see Chapter 6).

Normal development of embryos over the implantation and immediate post-implantation period has proved much more difficult to achieve *in vitro*, but a technique for culturing mouse embryos from the blastocyst to the mid-somite, beating-heart stage has been developed by Y.-C. Hsu in Baltimore. It requires the medium to be supplemented for the first 2 days of culture with fetal calf serum, and after that with human umbilical cord serum.

Denis New of Cambridge has shown that once implantation in the uterus has been completed and gastrulation has been initiated, embryos (at least in the mouse, rat and hamster) can be dissected out and grown in circulating medium or on plasma clots up to the late somite stage. The presence of serum in the medium seems essential.

Twins

Twins can be monozygotic (one-egg, identical) or dizygotic (two-egg, fraternal). Dizygotic twins are formed when two eggs are shed in a single ovulation period, and fertilized by two separate spermatozoa. The resulting young resemble each other genetically no more than do any other brothers and sisters. Indeed, the difference in weight between male and female lambs is greater in male–female twin pairs, because of embryonic competition, than when male and female lambs are born in like-sexed twin pairs. In species that normally produce several young at a time and are therefore termed polytocous (litter-bearing), the concept of dizygotic twinning scarcely applies.

Injection of gonadotrophic hormones or ovulation-inducing drugs like clomiphene can induce multiple ovulations and hence promote dizygotic twinning. This technique finds practical application in cattle breeding, while in women the birth of triplets, quadruplets, quintuplets or sextuplets occasionally occurs as an undesirable side-effect of the use of these compounds to combat infertility. The incidence of spontaneous dizygotic twinning in man varies widely among different populations, as it is affected not only by genetic but also by dietary and other environmental factors (see Table 1.2).

When two or more embryos are implanted in the same uterus, they may share a common or conjoined blood circulation (placental anastomosis), with the consequence that hormones and circulating cells are transferred between embryos. In cattle, when twinning occurs it is almost always accompanied by placental anastomosis; twin calves therefore share each other's blood groups and can accept skin grafts from one another and if they are of opposite sex, the sexual development of the female (freemartin) is impaired (see Chapter 3). In contrast, marmosets almost invariably produce dizygotic twins and, although exchange of blood cells regularly occurs, no abnormalities of sexual development have been observed.

Monozygotic twins originate from a single fertilized egg, and hence resemble one another very closely in all genetically determined characters. For instance, they are nearly always of the same sex – the only known exceptions are a few cases reported in our own species where loss of a sex chromosome from one twin early in development has led to the birth of XO (sterile female) and XY (normal male) 'identical' twins (see Chapter 3). The incidence of monozygotic twinning is much lower than that of dizygotic, and varies very little among different populations (Table 1.2). In cattle, up to 5 per cent of all births (depending on breed) are twins, but only about 0.1 per cent of births are monozygotic twins.

Many instances of monozygotic twinning probably originate after implantation. A single blastocyst implants, and the single inner cell mass then forms two primitive streaks, giving rise to two separate individuals. In man, the type of placentation provides valuable clues as to the stage at which the twinning originated (see Chapter 2).

If monozygotic twinning is possible then it is not surprising that the early

Table 1.2. *Geographical variations in the incidence of monozygotic and dizygotic twinning in women*

Country	Incidence per 1000 births		
	Total	Dizygotic	Monozygotic
Japan	5.6	2.7	2.9
Spain	9.1	5.9	3.2
France	10.8	7.1	3.7
India	11.1	6.8	4.3
USA 'White'	11.3	7.1	4.2
West Germany	11.5	8.2	3.3
Sweden	11.8	8.6	3.2
England and Wales	12.3	8.8	3.5
Greece	13.8	10.9	2.9
USA Negro	15.8	11.1	4.7
Congo	22.0	19.0	3.0
Nigeria	46.0	42.0	4.0

Source: From I. MacGillivray, P. P. S. Nylander and G. Corney. *Human Multiple Reproduction.* Saunders; Philadelphia (1975).

Fig. 1.9. (*a*) Early pregnancy
in a nine-banded armadillo
Dasypus novemcinctus,
showing the four identical
embryos derived from a
single fertilized egg. The
chorionic sac has been
opened and the embryos
are viewed from the dorsal
position, each lying within
its own amniotic sac.
(From K. Benirschke. *Fetal
Homeostasis*, ed. R. M.
Wynn. *N.Y. Acad. Sci.* **1**
237 (1965).) (*b*) A later
stage of pregnancy in the
nine-banded armadillo,
showing the identical
quadruplet fetuses, separately
attached to the common
placenta. (Courtesy of K.
Benirschke.)

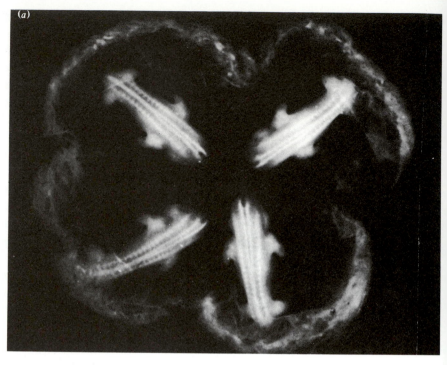

(*a*)

(*b*)

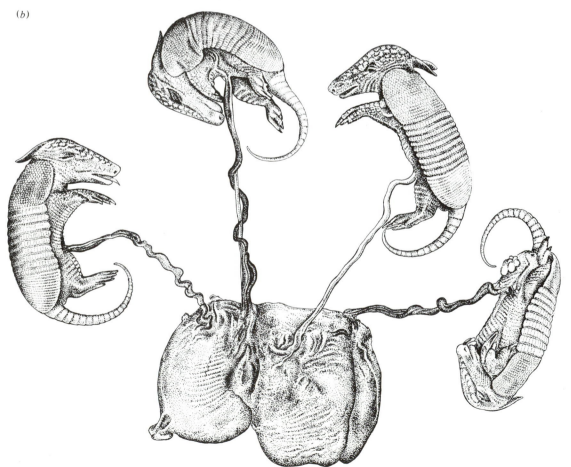

embryo can also divide itself into more than two parts, an outcome known as polyembryony. The classic example of monozygotic polyembryony is seen in the nine-banded armadillo, where the single blastocyst invariably produces four primitive streaks, so that the pregnancy results in the birth of four monozygotic young (Fig. 1.9). In the twelve-banded armadillo there can be as many as eight monozygotic young.

Embryos in the reproductive tract

After fertilization in the ampullary region of the oviduct or Fallopian tube, the mammalian egg is transported into the uterus (Fig. 1.10). In most marsupials this may take 24 hours or less, while the embryos of some carnivores may stay in the oviduct for up to a week, but in most mammals the journey from ampulla to uterus takes 2–4 days (Table 1.3). The rate of transport down the oviduct is not uniform: in some species the embryos remain for a long time at the ampullary–isthmic junction, halfway along the oviduct, while in others they are held up at the entrance to the uterus, the utero-tubal junction (see Book 1, Chapter 5).

In rodents, it seems essential for the survival of the embryo that the

Fig. 1.10. Development of the human embryo in the reproductive tract, from fertilization to implantation. (After H. Tuchmann-Duplessis, G. David and P. Haegel. *Illustrated Human Embryology*, vol 1. Springer-Verlag, New York; Chapman and Hall, London; Masson et Cie, Paris. (1971).)

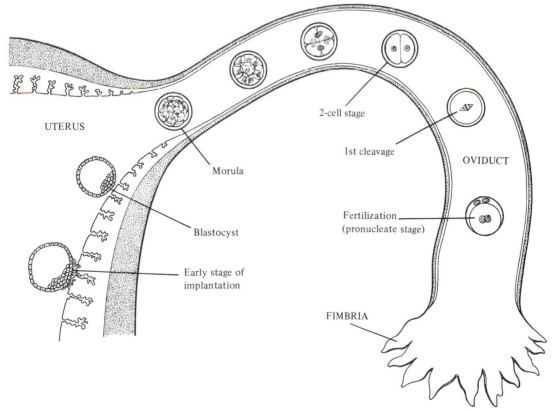

2-cell stage

UTERUS

1st cleavage

OVIDUCT

Morula

Fertilization
(pronucleate stage)

Blastocyst

Early stage of
implantation

FIMBRIA

The embryo

Table 1.3. *Timing of gestation*

Animal	2-cell	4-cell	16-cell	Blastocyst	Entry into uterus (days)	Implantation (days)	Total gestation (fertilization to birth) (days)
Man[a]	1½ days	2 days	3 days	4 days	2–3	8–13	252–74
Rat[a]	1–2 days	2–3 days	4 days	4½ days	3	5	20–22
Mouse[a]	21–3 h	38–50 h	60–70 h	66–82 h	3	4	19–20
Rabbit[a]	21–5 h	25–32 h	40–7 h	75–96 h	2.5–4	7–8	30–32
Guinea pig[b]	23–48 h	30–75 h	107 h	115 h	3.5	6	63–70
Rhesus monkey[a]	26–49 h	24–52 h	4–6 days	7–8 days	3	9–11	159–74
Horse[a]	24 h	30–6 h	98–100 h	6 days	4–5	28	335–45
Cow[a]	24 h	2–3 days	4 days	7–8 days	3–4	30–5	275–90
Sheep[a]	24 h	42 h	3 days	6–7 days	2–4	15–16	145–55
Pig[a]	16–20 h	25–74 h	80–120 h	5–6 days	2–2½	11	112–15
Ferret[b]	51–71 h	64–74 h	95–120 h	4½–6 days	5–6	7–8	42
Mink[b]	3 days	3–4 days	5–6 days	8–10 days	8	25	42–52
Cat[b]	40–50 h	3 days	4 days	5–6 days	4–8	13–14	52–65

[a] Times from ovulation.
[b] Times from coitus (first coitus with mink).

oviduct should not be traversed too rapidly. At oestrus, the distended uterus provides a favourable environment for spermatozoa, but a very unfavourable one for eggs or embryos; not until at least a day later does the uterine environment become tolerable for the embryo. The zona has been shown to be essential for the transport of cleavage stages in the mouse, although blastocysts experimentally injected into the top of the oviduct are transported to the uterus whether or not the zona pellucida is present. We surmise that an important function of the zona pellucida is to hold the cleaving embryo together during its passage down the oviduct and to prevent it sticking to the oviduct walls.

Once the embryos have entered the uterus, a period of at least a day (in mouse or rat) and several weeks in the horse intervenes before implantation begins. Muscular action is important in animals with more than one young per litter (polytocous species), in order to spread the embryos along the length of the uterus, and thus minimize crowding, which in later pregnancy might lead to embryonic death. Philip Dziuk and his colleagues at Urbana have examined in detail the timing and distribution

Fig. 1.11. Diagram to represent the result of transferring embryos of a black breed of pigs to the left uterine horn of a sow, and eggs of a white breed to the right uterine horn. When the sow was killed at the 90th day of gestation, the extent of internal migration was readily apparent from the colour of the fetuses. (From A. W. Marrable. *The Embryonic Pig*, p. 102. Pitman Medical; New York (1971).)

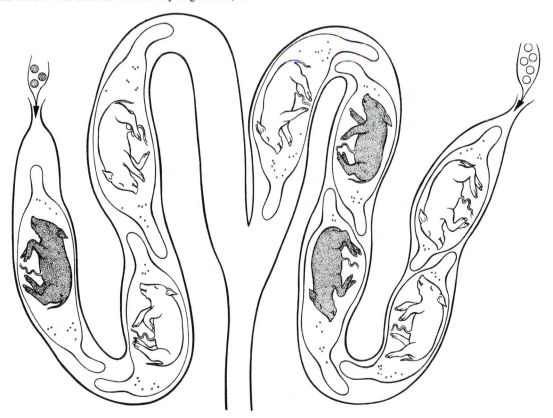

of embryos entering the pig uterus from one side only, and find that the proportion of the total uterus reached by such embryos rises from 13 per cent on the sixth day to 86 per cent on the twelfth day (see Chapter 2).

Where the two uterine horns open separately into the vagina, as in the rat, eggs cannot migrate internally from one uterine horn to the other. 'External migration' – the passage of an unfertilized egg from the ovary on one side across the peritoneal cavity and into the top of the oviduct on the other side – can occur in species lacking a closed ovarian capsule. In the mouse, internal migration of embryos can occur between the two uterine horns, but in undisturbed pregnancies is rare, affecting less than 1 per cent of embryos. In the pig, on the other hand, it occurs very frequently. In an elegant experiment, again performed by Philip Dziuk's group, eggs from a black breed of pigs were transferred into the left uterine horn and eggs from a white breed into the right horn of a sow. When the sow was killed at the 90th day of gestation, the distribution of black and white fetuses shows the extent of internal migration of embryos between the two uterine horns (see Fig. 1.11). From data on the relative distributions of corpora lutea and implantation sites between the two sides of the reproductive tract, it has been calculated that an egg shed from one ovary has an equal probability of implanting in the uterine horn on the opposite side as on the same side of the tract. In consequence, although the numbers of eggs shed from the two ovaries are negatively correlated, the numbers of embryos implanted in the two horns show a strong positive correlation, and indeed seldom differ by more than one. In intersex pigs, which have a functional ovary on one side and a testis or ovotestis on the other, all the eggs are shed on the ovarian side, but most of the implantation sites are found in the uterine horn on the opposite side. This suggests that the testicular tissue exerts a local hormonal effect on the uterine horn of that side, interfering with the normal pattern of migration of embryos between sides.

Transuterine migration is also common in sheep, but curiously it seldom occurs in cattle, so that problems arise if two fertilized eggs are transferred to one uterine horn in embryo transfer experiments; one of the embryos usually dies. In some other ungulate species, like the impala, pregnancies almost always occur in the right uterine horn of an otherwise symmetrical reproductive tract, even though the ovulations may often have occurred in the left ovary.

In the mouse, the blastocysts are not only spread down the length of the uterine horns by the muscular action of the uterus, but are positioned by the same process along the antimesometrial surface of the endometrium, i.e. on the wall opposite to that by which the uterus is suspended in the peritoneal cavity. The role of the blastocyst must be a passive one, since beads or small pieces of muscle placed in the uterus are treated in the same way.

In mammals, the fertilized egg is preloaded with RNA and proteins, which play an essential role during early development. But the embryo's own genes are quick to get into action, and throughout the whole of pre-implantation development, the mammalian embryo remains relatively independent of the maternal environment. This fortunate circumstance provides experimental embryologists with a great opportunity to unravel the causal basis of the first stages of differentiation. From implantation onwards, the situation is very different: the embryo is almost entirely at the mercy of the mother, and perishes swiftly if her hormonal, physiological and immunological adaptations to pregnancy are inadequate.

Suggested further reading

Nuclear transplantation in *Mus musculus*: developmental potential of nuclei from preimplantation embryos. K. Illmensee and P. C. Hoppe. *Cell*, **23**, 9–18 (1981).

Sequential X-chromosome inactivation coupled with cellular differentiation in early mouse embryos. M. Monk and M. Harper. *Nature, London*, **281**, 311–313 (1979).

A method for culture of micromanipulated sheep embryos and its use to produce monozygotic twins. S. M. Willadsen. *Nature, London*, **277**, 298–300 (1979).

Physiological aspects of growth and development of the preimplantation mammalian embryo. J. D. Biggers and R. M. Borland. *Annual Review of Physiology*, **38**, 95–119 (1976).

The generation and recognition of positional information in the preimplantation mouse embryo. M. H. Johnson, H. P. M. Pratt and A. H. Handyside. In *Cellular and Molecular Aspects of Implantation*. Ed. S. Glasser and D. Bullock. Plenum Press; London (1981).

Genetics of the early mouse embryo. A. McLaren. *Annual Review of Genetics*, **10**, 361–388 (1976).

Parthenogenesis in mammals. D. G. Whittingham. In *Oxford Reviews of Reproductive Biology*, vol. 2. Ed. C. A. Finn, Oxford University Press (1980).

The Mammalian Egg. C. R. Austin. Blackwell Scientific Publications; Oxford (1961).

Embryogenesis in Mammals. Ed. K. Elliott and M. O'Connor. Ciba Foundation Symposium. Elsevier; Amsterdam (1976).

Mammalian Chimaeras. A. McLaren. Cambridge University Press; London (1976).

Preimplantation Stages of Pregnancy. Ed. G. E. W. Wolstenholme and M. O'Connor. Ciba Foundation Symposium. Churchill; London (1965).

2

Implantation and placentation

M. B. RENFREE

One of the most fascinating aspects of biology is the conception, growth and birth of the young. Without some understanding of the way in which a genetically dissimilar fetus taps the mother's nutrients and directs her endocrine system, we cannot say that we understand reproductive biology, which is really the science of the transmission of life.

In mammals, the fertilized egg proceeds through the cleavage stages in a relatively autonomous manner, up to the formation of the blastocyst (Chapter 1). The blastocysts of marsupials (unlike those of eutherian mammals) have no inner cell mass, but both eutherian and marsupial mammals are endowed with an outer cell layer – the trophoblast – which will eventually form their placentae (Fig. 2.1). The duration of the pre-implantation stage varies between species, ranging from about 4–6 days in rabbits and mice, through nearly 20 days in kangaroos and wallabies, to 15–30 days in sheep and cattle. The pre-attachment phase

Fig. 2.1. Diapause blastocysts of (*a*) a bat *Miniopterus schreibersii*, and (*b*) a wallaby *Macropus eugenii*. The bat blastocyst has already lost its zona pellucida, and in this histological section the inner cell mass (ICM) can be clearly distinguished from the trophoblast (T). The wallaby blastocyst is intact, as flushed from the uterus. The zona pellucida (ZP) is very thin and difficult to see, but there is a broad mucoid coat (MC) separating it from the shell membrane (SM) (see also Fig. 2.2). There is no inner cell mass, and in the surface view of the whole blastocyst the trophoblast can be distinguished. (From M. B. Renfree. In *Mechanisms of Dormancy and Developmental Arrest*, ed. M. E. Clutter. Academic Press; New York (1978).)

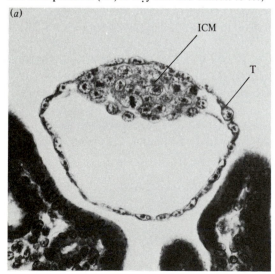

(*a*)

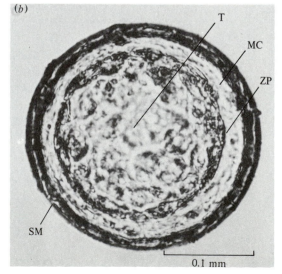

(*b*)

0.1 mm

may be extended for even greater periods by the suspension of blastocyst development, usually when the blastocyst has grown to around 100 cells, a phenomenon known as embryonic diapause. Diapause may last as long as one year in some marsupials, or 10 months in the European badger *Meles meles*, or as little as 10 days in the lactating rat or mouse. Transformation from the fertilized egg to the blastocyst stage can take place *in vitro*, so that growth during the pre-implantation period is not necessarily dependent on the uterus, but the transition from blastocyst to implanted and expanding embryo requires a receptive uterus and a physiological trigger to initiate the transition. During implantation and for some time thereafter the embryonic membranes form and develop into a placenta, which makes efficient nutrient exchange between mother and embryo possible.

The control of implantation, and the control of embryonic diapause, depends on a complex interplay of physiological and environmental mechanisms, regulating growth of the blastocyst and ultimately the formation of a wide variety of different types of placental attachment. This chapter will describe the events from blastocyst activation and growth to the formation of the placental membranes in a range of mammalian species.

Implantation

Once the blastocyst stage has been reached, a sequence of changes begins in the uterus and in the embryo that finally results in implantation. The extent of implantation varies according to the type of placenta developed. The precise point in time at which implantation begins is difficult to define – early workers thought that it had begun when it was no longer possible to wash the blastocyst out of the uterine lumen by flushing through with saline. With the electron microscope, changes can be observed in the endometrium earlier than this. Around the time of implantation in many species there is an increased capillary permeability which can be readily detected by the injection of a vital dye such as Pontamine Sky Blue or Geigy Blue – the uterus of the mouse shows blue bands corresponding to each implantation site.

The early embryo is protected from leucocytic attack within its muco-polysaccharide coat, the zona pellucida, and within this envelope it is electrostatically a relatively uncharged object. Once the blastocyst has 'hatched' from the zona, it becomes negatively charged and very sticky, so that another function of the zona may be to prevent premature attachment of the embryo to the tubal (oviduct) epithelium (see also Chapter 1).

The zona pellucida is usually very obvious in eutherian blastocysts, but is quite thin and difficult to see in marsupial eggs which have an additional mucoprotein coat, like the rabbit egg, and a shell membrane (Figs 2.1 and 2.2). However, all mammalian embryos need to shed the zona in order to

Table 2.1. *Characteristics of embryonic diapause in different groups of mammals*

State of blastocyst	State of corpus luteum	Type of delay	
		Facultative (e.g. lactational)	Obligatory (e.g. seasonal)
1. No growth Unilaminar blastocyst with zona pellucida	Quiescent	Wallabies and kangaroos	Some wallabies
2. No growth Unilaminar blastocyst without zona pellucida	Active	Laboratory mouse and rat	
3. Some growth Unilaminar blastocyst with zona pellucida	Quiescent Some activity		Mustelids Bears and seals

Fig. 2.2. Cross-sectional views of mammalian blastocysts showing external coats. Sub ZL, sub-zona layer; other abbreviations as in Fig. 2. 1. (Redrawn from Wimsatt (1975) – see Suggested Further Reading.)

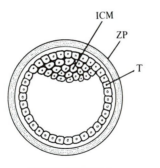

MOST EUTHERIA

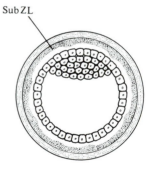

SEALS

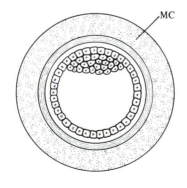

RABBIT

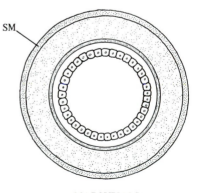

MARSUPIALS

be free to adhere to the uterine epithelium. Loss of the zona is not necessarily followed immediately by implantation; in some diapausing species such as the laboratory rat and mouse, the roe deer and the armadillo, the blastocyst remains in suspended development after hatching from the zona (Table 2.1).

In the mouse, this breakdown or disruption of the zona occurs in an oestrogen-sensitized uterus, either through the action of a uterine protease, or as a result of altered pH. Some protease may also come from the embryo's trophoblast cells. In the rabbit, the development of protease activity is dependent on progesterone. Once hatching has been achieved, the trophoblast begins a rapid phase of growth, resulting in either an expansion of the size of the blastocyst prior to attachment, or an immediate invasion of the maternal uterine epithelium.

Types of implantation

In species in which the blastocyst expands somewhat before attachment, such as the rabbit, dog, ferret, and many marsupials, the trophoblast tissue has a large area exposed to the uterine lumen, giving a *centric* implantation.

Fig. 2.3. Common types of implantation patterns observed in mammals: centric, eccentric, partly, or completely interstitial. Although implantation may also be mesometrial or lateral, it is most commonly antimesometrial, as shown in these diagrams. Abbreviations as in Fig. 2.1. (Redrawn from Wimsatt (1975) – see Suggested Further Reading).

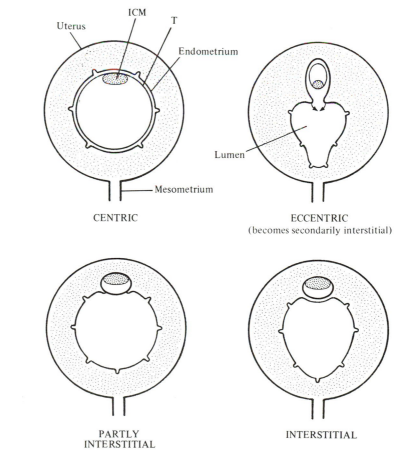

CENTRIC

ECCENTRIC
(becomes secondarily interstitial)

PARTLY
INTERSTITIAL

INTERSTITIAL

Those with small, unexpanded blastocysts, such as the mouse, rat and hamster, may form an implantation chamber within the epithelium of the uterus to produce an *eccentric* implantation. Other species, such as the guinea pig, lemming, chimpanzee and human, have small blastocysts which penetrate the epithelium and pass into the sub-epithelial connective tissue to give an *interstitial* implantation (Fig. 2.3).

Where the blastocyst is small, and implantation is eccentric or interstitial, the implantation site bears a regular relationship to the mesometrium – the connective tissue that attaches the uterus to the body wall, and through which run numerous blood vessels. Thus in the mouse and rat, implantation is antimesometrial (opposite to the mesometrial attachment), while it is mesometrial in certain bats and the tarsier. In the Artiodactyla (cloven-hoofed ungulates), whose blastocysts elongate quite dramatically before attachment (Fig. 2.4), implantation occurs at predetermined sites, the endometrial caruncles. In the horse, pig, kangaroo and wallaby there are no such special areas, and attachment occurs at the unspecialized endometrial surface and over the entire trophoblastic surface.

The mechanisms that control the sites of implantation are not clearly

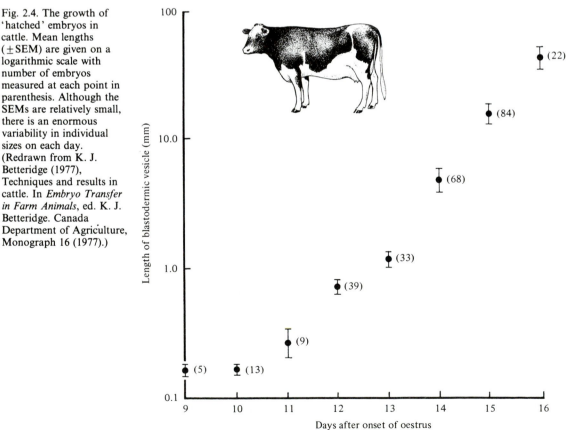

Fig. 2.4. The growth of 'hatched' embryos in cattle. Mean lengths ($\pm$ SEM) are given on a logarithmic scale with number of embryos measured at each point in parenthesis. Although the SEMs are relatively small, there is an enormous variability in individual sizes on each day. (Redrawn from K. J. Betteridge (1977), Techniques and results in cattle. In *Embryo Transfer in Farm Animals*, ed. K. J. Betteridge. Canada Department of Agriculture, Monograph 16 (1977).)

understood; for example in rodents there are no clear morphological differences between the mesometrial and antimesometrial mucosa. However, if small pieces of muscle or tumour tissue are placed in a receptive mouse uterus, the tissue always attaches to the antimesometrial site. Likewise, oil injection, which stimulates a decidual reaction in the mouse, always does so antimesometrially.

Blastocysts tend to be spaced out along the length of the uterus – an important adaptation in polytocous species (which have several embryos). Harland Mossman in 1937 thought that this was due to a 'refractory zone' round each blastocyst, but Anne McLaren and Donald Michie challenged this interpretation as they found that uterine contractions were important in mixing and spacing embryos. In some species blastocysts travel from one uterine horn to the other (see Chapter 1).

Changes in the endometrium during implantation

The purpose of implantation is to bring the blood vessels of the embryo into functional communication with the maternal blood supply. As we have seen (Fig. 2.3) the blastocyst therefore comes to occupy a position either against the uterine epithelium, or at the other extreme it embeds itself in the uterine stroma. In those species in which the blastocyst merely rests against the walls of the uterus, the endometrium 'closes down' on the blastocyst. However, this endometrial reaction also occurs in pseudo-pregnancy and in some delayed implanters, so it is not strictly a part of implantation. During the re-attachment phase, the microvilli of the uterus and blastocyst interlock. The attachment or adhesion phase that follows is the result of a flattening of the opposed surfaces. The microvilli disappear, and are replaced by club-shaped projections, possibly because of an increased absorptive capacity of the uterine epithelium. Similar changes can be induced in the endometrium by inserting a plastic bead into the uterus of a progesterone-primed oestrogen-injected mouse. In those species like the pig and wallaby that have only a superficial attachment, the interlocking microvilli represent the ultimate placental union.

At the other extreme are those species like the human whose blastocysts actively invade the endometrium and eventually make contact with the stroma (Fig. 2.5). This is due not only to invasive properties of the blastocyst, or more correctly the trophoblast layer, but also to uterine epithelial cell degeneration. The passage of the blastocyst into the stroma involves the breaking of junctional complexes, and it has been suggested that the trophoblast secretes proteolytic enzymes to allow this to take place: lysosomes may also be involved in phagocytosis of the stromal cells.

The increased permeability of the blood vessels at the implantation site is followed by oedema or swelling of the stroma surrounding the blastocyst. Associated with this are changes in the stromal collagen: thick fibres are replaced by fine fibres. Mitotic divisions are seen in the stromal cells, and those in contact with the blastocyst become arranged in rows. This is the first stage in the differentiation of stromal cells into decidual cells, so called because they are sometimes shed at parturition.

The decidual cell reaction. The decidual cell reaction is confined to rodents, higher primates and insectivores, and although some changes are seen in the uterus, true decidual cells do not occur in carnivores or artiodactyls. The decidual cells consist of at least three transformed stromal cell types which differ morphologically from each other, depending on their position in the uterus. The decidual cells are large and have more cytoplasmic inclusions than do unstimulated stromal cells. Decidual cells contain

Fig. 2.5. (*a*) Section of the embryo of a rhesus monkey adhering to the uterus at day 9 after conception, shortly after the initiation of implanta-tion. The layer of vertically aligned cells on which the (hollow) embryo rests is the endometrial epithelium, the lining of the uterine cavity; below the endometrium is the decidual tissue. In the two areas where the embryo and uterus adjoin, the trophoblast has begun to invade the endometrium. × 400. (*b*) An early human implantation site, at a slightly later stage than that of the rhesus monkey in (*a*), showing the invasion of the endometrium by trophoblast which has now apparently become syn-cytial (s). ((*a*) From A. E. Beer and R. Billingham, *Sci. Amer.* **230**, 36–46 (1974); and (*b*) from A. C. Enders, D. J. Chavez and S. Schlafke. In *Cellular and Molecular Aspects of Implantation*, ed. S. R. Glasser and D. W. Bullock. Plenum Press; New York (1981).)

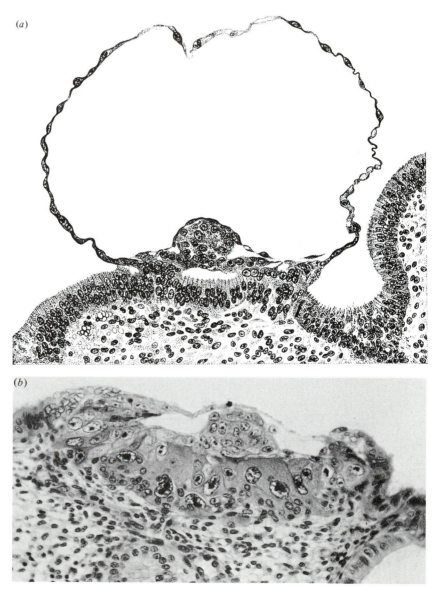

glycogen granules, which are eventually lost, and lysosomes, which may be involved in the autolysis of the decidual cells during the process of implantation. In women, the initial decidual transformation occurs in the stromal cells throughout the uterus late in the luteal phase of the menstrual cycle, but in the majority of species the transformation is confined to areas adjacent to a blastocyst, and is dependent on a stimulus from the blastocyst.

What is the nature of the decidual stimulus? A variety of agents can initiate the development of decidual cells such as oil, air or saline, injected into the uterine lumen, the intraperitoneal injection of pyrathiazine, or the introduction of sea urchin eggs, fixed rat or mouse ova, or small pieces of agar into the uterine lumen. Glass beads, however, are not effective. There are a number of theories to explain the decidual cell response. Histamine released from the endometrial mast cells may initiate decidual transformation, but the nature of the stimulus that causes this release is unclear. Recent experiments with histamine H_1 and H_2 antagonists have shown that these inhibit both the increased capillary permeability and implantation. The response to air may be due to its concentration of CO_2: the idea is that in a normal pregnancy CO_2 derived from blastocyst respiration will trigger the decidual cell response. Since the initial reaction between endometrium and blastocyst still occurs in the presence of actinomycin D, an inhibitor of DNA transcription, the signal does not apparently involve the transfer of newly formed RNA from the nucleus.

The decidual cells have a finite life, and progesterone is essential to maintain them. Progesterone acts as a permissive hormone in this case, and is not involved in the morphogenesis of the response (Fig. 2.6). In normal pregnancy, the dead decidual cells are phagocytosed by the trophoblast, which has led to the suggestion that decidual degeneration could provide a source of nutrition for the embryo.

Fig. 2.6. (*a*) Levels of cyclic AMP (cAMP) in the uterus after stimulation of the decidual cell response by oil. Ovariectomized mice were primed with oestrogen and progesterone and the right uterine horn (●) stimulated by intraluminal injection of sesame oil. Mice (*n* = 6) were killed each hour after injection, and stimulated (●) and non-stimulated (○) horns were assayed individually for cAMP. Values are means. (Redrawn from J. C. Rankin, B. E. Ledford and B. Baggett. In *Cellular and Molecular Aspects of Implantation*, ed. S. R. Glasser and D. W. Bullock. Plenum Press; New York (1981).) (*b*) Growth and regression of uterine horn weights (●) and decidual tissue weights (○) of a group of oil-injected ovariectomized mice sensitized with oestrogen and progesterone. In spite of continued injection of progesterone, regression of the decidua occurred after 6 days. (Uterine horn weights from C. A. Villee, E. G. Armstrong, D. J. Talley and H. Hoshiai. In *Molecular and Cellular Aspects of Implantation*, p. 242, ed. S. R. Glasser and D. W. Bullock. Plenum Press; New York (1981). Decidual tissue weights from C. A. Finn. In *The Uterus*, p. 83, ed. D. G. Porter and C. A. Finn. Elek Press; London (1975).)

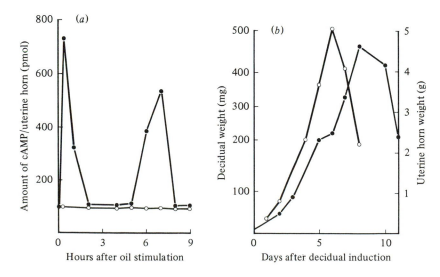

A variety of other factors have also been suggested as initiators of the decidual cell response, including prostaglandins and cyclic nucleotides (Fig. 2.6). Since decidual cells are found in only a few mammalian orders, they probably represent a special development to contain and regulate growth of the trophoblast in species with interstitial implantation.

Delays in gestation and embryonic diapause

The length of gestation is generally proportional to the size of the young and its state of maturity at birth. Although the gestation length is usually fixed for a given species, there are mechanisms by which it can be varied according to need.

'Delayed fertilization' is a rare phenomenon found in a few species of bat. Characteristically, spermatozoa from a mating in autumn remain viable in the oviduct or uterus near the tubo-uterine junction until the female bat ovulates in the spring; fertilization then occurs and development promptly follows. Conversely, in 'retarded development', which also occurs only in two genera of bat, implantation occurs shortly after the blastocyst appears in the uterus, but the formation of the placenta is slow compared to closely related species, and the early stages of embryogenesis are retarded. In another bat *Pipistrellus pipistrellus*, pregnancy can be extended by the entry into torpor of the pregnant mother in response to low ambient temperature. But by far the commonest device for regulating gestation length is embryonic diapause, which so far has been described in almost 90 species of mammal. In most cases, it seems that the fate of the blastocyst is controlled proximately by the uterus, which itself is subject to endocrine control originating ultimately from other physiological or seasonal environmental factors (Table 2.1). Since the control of embryonic diapause offers an interesting way of studying the events leading up to implantation, we will consider diapause in some detail in discussing the control of implantation.

Control of implantation

Environmental control of embryonic diapause

The influence of season on the breeding of mammals is well documented, and is dealt with more fully elsewhere in these volumes (Book 3, Chapter 3 and Book 4, Chapters 2 and 4 of the Second Edition). In most species, the timing of implantation is set by the time of copulation and is not directly influenced by season, whereas environmental factors can control the timing of ovulation, but in those species that have embryonic diapause, environmental cues may also influence the time of blastocyst activation and subsequent implantation.

Premature activation of the diapausing blastocyst and implantation have been induced in the European badger and mink by altering the daily photoperiod. Embryos of the mink, long-tailed weasel, Western spotted skunk and pine marten (all mustelids) normally implant just after the shortest day, whereas embryos of the European badger implant after the longest day (Fig. 2.7). Temperature may also influence the timing. In Idaho, Rodney Mead and colleagues found that in the Western spotted skunk, blinding or constant darkness prolonged diapause but surprisingly, pinealectomy had little or no effect and did not induce implantation. It appears that in this species light, acting via the eyes, stimulates the hypothalamo-pituitary axis to increased secretion of gonadotrophins which then stimulates luteal activity, which in turn induces uterine secretion and thus blastocyst activation. Curiously, the Eastern spotted skunk, a closely related species, never shows embryonic diapause at all. Although we have yet to understand how this difference evolved, it must certainly be a powerful species-isolating mechanism as it effectively prevents interbreeding of the two contiguous populations.

The tammar wallaby has a remarkably synchronized reactivation period: blastocysts resume development on or about the summer solstice (December 22 in the Southern hemisphere), and most tammars in the wild are carrying newborn pouch young about a month later (Fig. 2.8). Experimental alteration of photoperiod can induce premature implantation in tammars, as Hugh Tyndale-Biscoe, Dick Sadleir and John Hearn showed in Canberra. In this species, several experiments suggest that the pineal is the mediator of the photoperiodic changes that ultimately result in reactivation of the corpus luteum and the blastocyst. Bob Seamark and

Fig. 2.7. Two newborn badger cubs, 8 days old, born at the normal time in February 1980, and a 6-month-old badger born in September, 1979, after its mother had been subjected to an artificial winter in a climate-controlled chamber. The mothers of these differently aged young were all mated at the same time in February, 1979. (Redrawn from photograph by R. Canivenc and M. Bonnin, *J. Reprod. Fert., Suppl.* **29**, 27 (1981).)

Fig. 2.8. Patterns of reproduction in (*a*) the tammar
wallaby, a seasonal breeder,
(gestation 27 days), (*b*) the
agile wallaby, a continuous
breeder, (gestation 27
days), and (*c*) the red
kangaroo, an opportunistic
breeder (gestation 33 days).
A, activation; B, birth; E,
exit from the pouch; O,
oestrus and ovulation. The
successive stages of postnatal development are: I
and II, early and late
pouch young which suck
continuously, and III, the
'young-at-foot' which still
revisits the pouch to suck.
In a prolonged drought,
the young die in the reverse
order; eventually ovulation
is inhibited and the adult
female goes into anoestrus.
Arrowhead shows new
ovulation after return to
oestrus following drought-
breaking rain. ((*a*)
Redrawn from M. B.
Renfree, D. W. Lincoln,
O. F. X. Almeida and
R. V. Short, *Nature, Lond.*
293, 138–9 (1981); (*b*)
redrawn from D. W. Lincoln
and M. B. Renfree, *J.
Reprod. Fert.* **63**, 193–203
(1981).)

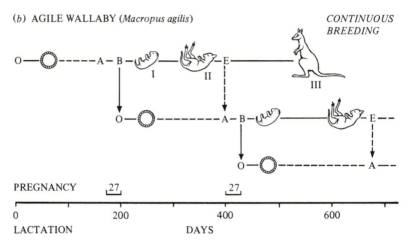

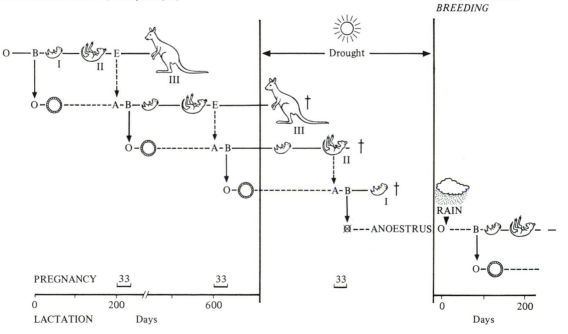

Dave Kennaway in Adelaide have shown that activity of the pineal enzyme hydroxy-indole-*O*-methyl-transferase, which converts serotonin to melatonin, drops on the longest day; in Perth we have shown that denervating the pineal by removal of the superior cervical ganglia eliminates the nocturnal rise in melatonin levels, and abolishes the seasonal diapause.

Nutrition may also influence diapause. In the quokka, a small wallaby, most diapausing blastocysts are lost within 6 months of their conception; the annual nutritional stress resulting from drought in late summer may be related to this loss of blastocyst viability. In lactating rats, implantation of their delayed blastocysts can be further delayed (causing loss of viability of the embryos) by reduction of food intake. The onset of seasonal rains can induce reactivation and implantation of the diapausing blastocysts in the equatorial African fruit bat *Eidolon helvum*. In the American fruit bat *Artibeus jamaicensis* the period of delay ensures that the young are born after the rains have begun in the spring, but the mechanisms by which this timing is entrained are as yet unknown. In the bent-winged bat *Miniopterus schreibersii* implantation of the diapause blastocysts is initiated only after the bats wake from hibernation in the spring, so that in this case environmental temperature may be the ultimate stimulus, and maternal body temperature and/or metabolic activity the proximate stimuli. However, once again, the precise mechanisms are unknown.

But the best example of a nutritional effect on diapause, often quoted erroneously as the 'typical' marsupial pattern, is that seen in two desert-living marsupials, the red kangaroo *Macropus rufus* and the euro or hill kangaroo *M. robustus*. These marsupials occupy marginal habitats, and Alan Newsome has shown that in good conditions in the Northern Territory the red kangaroo is a continuous breeder, with suckling of the pouch young the sole factor controlling diapause (Fig. 2.8). In favourable conditions they would also be supporting an older young-at-foot which sucks intermittently. However, if the quantity of available food declines, there is a heavy mortality of pouch young and subsequently of the young-at-foot. If poor conditions prevail, the diapausing blastocyst, like a reserve player, is called into action, resumes development and is born. But if the drought persists, it will survive in the pouch for only about 2 months, and a cycle of replacement of the pouch young every 6–8 weeks will continue until conditions improve after rain. After prolonged drought, many of the females enter anoestrus, but immediately after rain they return to oestrus, conceive and begin their reproductive cycle anew.

Endocrine control of embryonic diapause and implantation
Implantation is a two-way event, in which uterus and blastocyst interact with one another. Once the blastocyst stage has been reached, the embryo appears to require a physiological signal to trigger its expansion and subsequent implantation; the maternal hormonal stimuli are now relatively clearly defined in several species, particularly in the mouse.

The high levels of oestrogen at oestrus cause proliferation of the endometrium, and the progesterone subsequently produced by the newly formed corpus luteum renders the uterus receptive. This action of oestrogen is usually referred to as a priming action, which affects the timing and extent of stromal cell proliferation. Progesterone is crucial for at least the early stages of implantation, for in most species removal of the ovaries, or more specifically the corpora lutea, in early pregnancy causes failure of implantation (see Book 3, Chapter 7 of the Second Edition), whereas in most species administration of progesterone will maintain the pregnancy. An important exception is the armadillo, in which the diapausing embryos are actually stimulated to implant by ovariectomy.

In most species significant amounts of progesterone are not secreted by the newly formed corpus luteum until several days after ovulation. During this intervening period, hyperplasia of the uterine epithelium occurs, and this precedes the onset of the progesterone-induced phase of uterine glandular secretion. Progesterone secretion remains high during pregnancy in most eutherian species, although the main site of progesterone production may shift from the corpus luteum to the placenta.

In addition to luteal phase progesterone, there is evidence that oestrogen can also be crucial for implantation of the blastocyst. The quantity of oestrogen is critical, for implantation may be prevented by injecting high doses of oestrogen into pregnant animals. However, minute doses of oestrogen will terminate embryonic diapause and initiate implantation in rats and mice; conversely, blastocysts will remain in delay in ovariectomized rats and mice maintained on progesterone alone (Fig. 2.9). Rat blastocysts transferred to the uteri of ovariectomized recipients will only implant if the foster mothers have received progesterone 48 hours and oestrogen 24 hours before transfer.

In a number of species we can cause implantation in ovariectomized

Fig. 2.9. (*a*) The total amount of DNA polymerase activity in normal and diapausing mouse embryos. Normal embryos (●) were recovered from intact females at the times indicated. Diapausing (○) and activated (■) embryos were recovered from ovariectomized females treated with progesterone (○) or progesterone and oestradiol-17β (■). (*b*) The number of cells in normal and diapausing embryos. Embryos treated as in (*a*). (From H. M. Weitlauf and A. A. Kiessling. In *Cellular and Molecular Aspects of Implantation*, p. 130, ed. S. R. Glasser and D. W. Bullock. Plenum Press; New York (1981).)

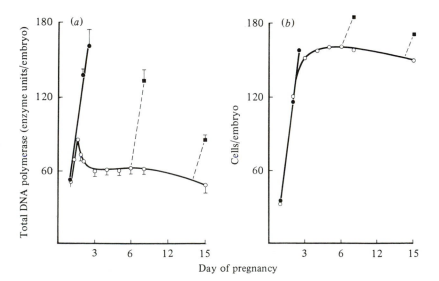

animals with progesterone alone, and in the rabbit, guinea pig, hamster, sheep, rhesus monkey and wallaby, oestrogen is apparently not necessary for implantation. Nevertheless, oestrogen is secreted during the luteal phase in several of these species, and in some it comes from the corpus luteum, so that its possible importance cannot be disregarded.

Luteal phase oestrogen seems to act primarily on the endometrium and not on the blastocyst itself, rendering the luminal surface of the epithelial cells 'sensitive' to implantation. This sensitivity is relatively short-lived, and thereafter the uterus becomes unreceptive to blastocysts. In some species, as exemplified by the pig, the main source of this oestrogen is the trophoblast (see Book 3, Chapter 7 of the Second Edition); rabbit blastocysts also contain significant quantities of oestrogen. However, the role of this trophoblastic oestrogen may be more to prevent luteolysis and so maintain progesterone production by the corpus luteum than to sensitize the endometrium.

In addition to steroids, some blastocysts also synthesize protein hormones. The human embryo starts to secrete human chorionic gonadotrophin (hCG), which is synthesized by the trophoblast, from about 8 days after ovulation, just after implantation, and this is responsible for transforming the corpus luteum of the menstrual cycle, which would otherwise regress after 14 days, into the maintained corpus luteum of pregnancy. Chorionic gonadotrophins have now been described in the rat, mouse, rabbit and pig, but in sheep and cows evidence for the production of gonadotrophin-like hormones in the peri-implantation period is weak (Fig. 2.10). In the horse, an unusual gonadotrophin, which has activities resembling follicular stimulating hormone and luteinizing hormone in other species but as yet no known hormonal activity in the horse itself, is secreted early in pregnancy by fetal allantochorionic cells, which invade the endometrium at implantation and form the endometrial cups – ulcer-like structures that form a ring around the pregnant horn of the uterus (see under 'The diffuse placenta', later in this chapter). Since this hormone is of fetal origin, Lief Wide suggests it should be called equine chorionic gonadotrophin (eCG) to bring it into line with the other chorionic gonadotrophins, rather than its original name, pregnant mares' serum gonadotrophin, or PMSG. We still do not know whether any steroid or protein hormones of trophoblastic origin have an essential role in the attachment of the blastocyst at implantation.

The endocrine requirements for reactivation and implantation of diapausing blastocysts are not well understood, but several detailed studies have been carried out. Bilateral ovariectomy of the European badger during diapause has no immediate effect on the blastocyst, but implantation cannot occur, presumably because the uterus is unable to develop the secretory endometrium that triggers the event. By contrast, ovariectomy early in delay in the nine-banded armadillo causes implantation in 18–20 days, whereas ovariectomy after implantation causes loss of the embryos.

Fig. 2.10. A summary of the embryo–maternal interactions that occur in mammals. There is a complex sequence of signals that passes among the ovary, uterus and blastocyst, beginning with ovulation and sensitization of the uterus, and ending with hatching, attaching and implantation of the blastocyst and maintenance of the corpus luteum by hormones of embryonic origin. Pituitary and hypothalamic hormones are not included in this diagram.

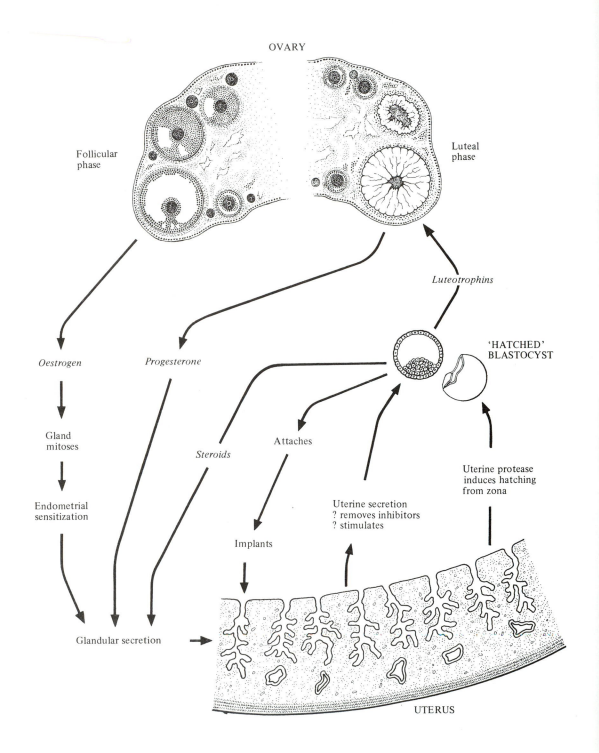

In at least two marsupials with delay, the quokka and the tammar wallaby, the ovaries are essential for the termination of diapause, but pregnancy will proceed to full term if ovariectomy or lutectomy is performed later in gestation. Hypophysectomy will even initiate blastocyst growth in the tammar, because this removes the source of prolactin which was responsible for keeping the corpus luteum quiescent, and hence maintaining the delay. In the Western spotted skunk, hypophysectomy prolongs the delay, and so gonadotrophins are apparently necessary in this species for the resumption of embryonic development. But all attempts to terminate diapause in mustelids – like the skunk, mink and badger – with exogenous gonadotrophins or steroids have been unsuccessful.

In rats and mice, lactational diapause can be terminated by an injection of a minute amount of oestrogen, or with very high doses of progesterone. In the red kangaroo and tammar wallaby, oestrogen will also initiate blastocyst development, although the embryos subsequently fail to develop. Progesterone alone will successfully reactivate diapausing blastocysts in

Fig. 2.11. Diagrammatic summary of the interrelated changes occurring in three wild species that have embryonic diapause: the spotted skunk (S), the roe deer (R), and the tammar wallaby (T). Arrows show the beginning of diapause (↑) and the resumption of development after diapause (↓). Diapause is characterized by relative stasis of all the parameters shown. However, reactivation of the diapause blastocyst occurs after hypophysectomy (*), in the tammar but not in the skunk. (From M. B. Renfree. In *Mechanisms of Dormancy and Developmental Arrest*, ed. M. E. Clutter, pp. 1–40. Academic Press; New York. (1981).)

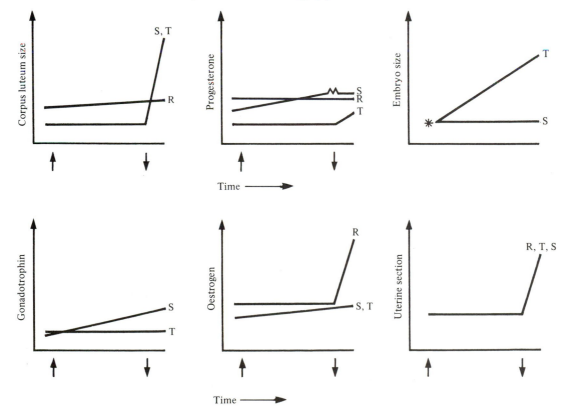

the red kangaroo, quokka and tammar, and about half will develop to full term in the tammar.

There seems to be no universal mechanism for initiating and terminating diapause, although several species may show certain common features. A schematic summary of the events controlling diapause is given for the roe deer, spotted skunk and tammar wallaby in Fig. 2.11.

Uterine responses

The various hormonal changes associated with the onset of pregnancy induce the uterus to become receptive and secretory. As we have seen, the period of receptivity is relatively brief, and from embryo transfer experiments we know that the arrival of the blastocyst in the uterus, or its reactivation after diapause, must be closely synchronized with uterine events if the pregnancy is to develop.

If mouse embryos are transferred to ectopic sites, such as the kidney, liver or brain, they will implant and grow almost normally. A mouse embryo will even develop in the testis of a rat – the wrong site, the wrong sex and the wrong species! Yet should an embryo be put into a non-sensitized uterus, implantation will fail. How, then, is the synchrony between embryo and uterus achieved? Zeev Dickmann and Vince De Feo transferred rat blastocysts in diapause to 'active' (i.e. hormonally stimulated) rat uteri, and found that the blastocysts resumed development. Conversely, 'active' blastocysts became dormant when placed into dormant uteri. Hugh Tyndale-Biscoe has also shown that diapause tammar wallaby and quokka embryos will develop following transfer to 'active' uteri. In cattle, sheep and pigs, with their long pre-attachment phases, there appears to be some tolerance to asynchrony, although the highest conception rates occur when the age of the donor's embryos is precisely synchronized with the stage of the recipient uterus. Tim Rowson and Bob Moor in Cambridge have demonstrated that cow and sheep embryos can tolerate asynchrony of ± 1–2 days.

Relatively little is known about biochemical changes in the uterine secretions that regulate embryonic development. Endometrial secretion is certainly influenced by the level of ovarian steroids, and administration of steroids can change the relative proportions of the proteins and volume of uterine secretions. There is as yet no definitive evidence that any single uterine protein has a crucial role in promoting either blastocyst growth or implantation, but overall composition of the uterine environment is undoubtedly important. Nevertheless, a great deal of effort has been devoted to characterizing and isolating a few uterine specific proteins.

In the rabbit, a uterine-specific protein was discovered at much the same time by Joe Daniel working in America who named it 'blastokinin' and by Henning Beier working in W. Germany who named it 'uteroglobin'; it appears in the uterine fluid in the peri-implantation period, and on the day of implantation comprises about 20 per cent of the total protein (Fig. 2.12).

Uteroglobin synthesis is stimulated by progestagens, but not oestrogens, and it also acts as a progesterone-binding protein. In the pig, maternally derived 'purple protein' accumulates in the allantoic fluid, and its synthesis is likewise stimulated by progesterone. If the sow's uterus is stimulated by progesterone or oestrogen, the chorionic sacs become longer and the allantoic fluid more copious; Fuller Bazer therefore suggests that uterine secretion may enhance placental development in pigs. In addition, the uterine proteins, many of which bind progesterone, may be important for delivery of progesterone to the blastocysts for further metabolism, especially in the pre-attachment phases. In mice, implantation is associated with an increased uterine protein content, but during diapause protein levels are low. In the roe deer and tammar wallaby, embryonic diapause is also associated with a deficiency of secretory material, and endometrial secretion reaches the highest levels at the time of embryonic reactivation and implantation.

Apart from proteins, uterine secretions contain carbohydrates, lipids, steroids and various ions. Potassium ions may be present in higher concentrations in uterine secretion than in serum, whilst zinc and calcium concentrations may vary before, during and after implantation. Any or all of these factors may influence blastocyst growth and placental formation.

Thus, although a specific role cannot yet be assigned to the uterine proteins, either individually or collectively, total uterine secretory activity

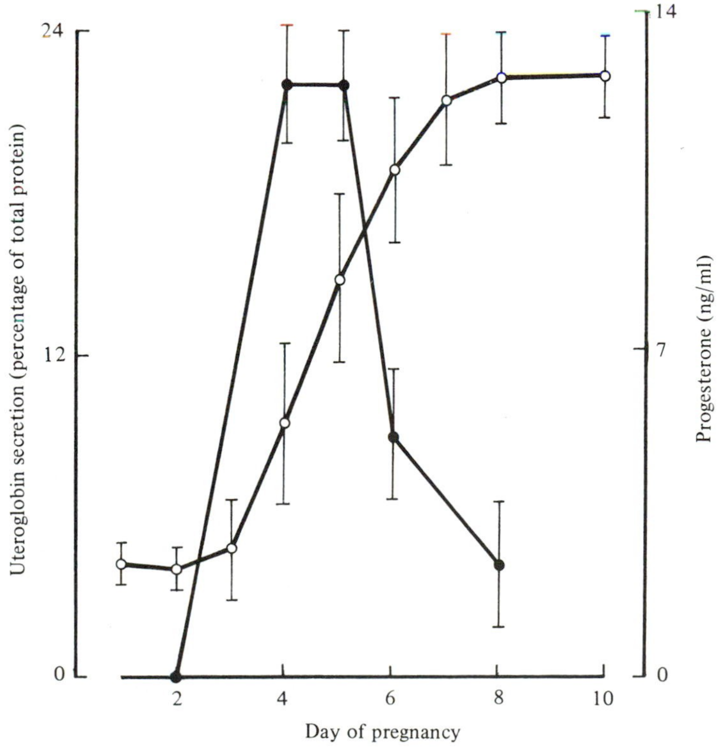

Fig. 2.12. Pattern of uteroglobin secretion as a percentage of total protein in uterine flushings (●) compared with concentrations of peripheral plasma progesterone (○) in early pregnancy in the rabbit. Vertical lines show means ± SEM. (From D. W. Bullock, L. W. L. Kao and C. E. Young. In *Cellular and Molecular Aspects of Implantation*, ed. S. R. Glasser and D. W. Bullock. Plenum Press; New York (1981).)

certainly seems to be important: it is under the influence of ovarian steroids and is a necessary pre-requisite for the developmental events leading up to implantation (Fig. 2.10).

The uterus as a site of embryonic growth

The uterus, consisting of an outer muscle layer, the myometrium, and an inner glandular layer, the endometrium, is a highly specialized organ. As we have seen, it is very responsive to changing hormone levels, and either the trophoblast and/or the uterus must have some unusual immunological properties, because the uterus does not reject the immunologically alien conceptus, although it is capable of rejecting other foreign tissues. This is a problem that has intrigued many biologists, but there is still no satisfactory explanation. Certainly the uterus *is* a rejection site, and is an organ in which transplantation antigens are readily detected and in which transplantation immunity is fully expressed; nevertheless, it will accept or reject tumours depending on its endocrine state. Furthermore, if dispersed

Fig. 2.13. Immunological relationship between mother and fetus involves the exchange of antigenic material in both directions. The probable site of fetal protection from a maternal immune response is the trophoblast, shown here in cross section of the mature placenta (enlarged). Although the trophoblast is in contact with maternal blood, it apparently does not evoke an effective immunological reaction, and may mask the fetal antigens. It may also repel maternal lymphocytes. (Redrawn from A. E. Beer and R. Billingham, *Sci. Amer.* **230**, 36–46 (1974).)

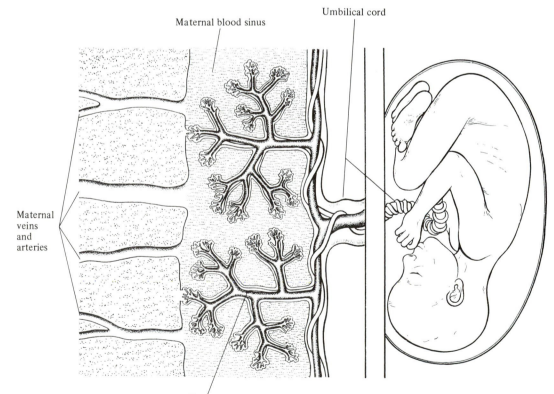

Maternal blood sinus

Umbilical cord

Maternal veins and arteries

Fetal veins and arteries in villus

cells from another individual of the same species (allograft) are injected into a uterine horn, subsequent grafts to that horn from the same donor are quickly rejected, although reproductive function is not impaired and embryos can still develop successfully in the same horn. So how is the embryo exempt from immunological attack?

As Alan Beer and Rupert Billingham have shown, pregnancy can be regarded as a graft–host relationship, but an unusual one in that the embryo seems indifferent to the maternal immune response. They believe that the continuous, unbroken layer of trophoblast cells acts as a dialysis membrane which is in direct, intimate contact with the maternal uterine tissue and blood (Fig. 2.13). Evidence seems to be emerging that the embryo establishes an immunological 'truce' with the mother, and either jams or controls her immune system so that the trophoblast enjoys an immunologically privileged status. The embryo probably does this in a number of ways.

In those species that form a decidua at implantation, the decidual tissue can give the early conceptus some immunological protection, although intrauterine skin allografts are rapidly rejected. The trophoblast itself is a privileged tissue, because it not only has fewer histocompatibility antigens on its surface, but trophoblast cells are coated with a physical barrier of sialomucin. In addition, the trophoblast tissue sheds surface antigens, which may reduce immunological recognition, and provide a stimulus for the production of maternal protective or 'blocking' antibodies that prevent the proliferation of antigenically reactive cells. Finally, there is evidence that the placental hormones may have immunosuppressive properties. Although there is great species variability of placental structure and intimacy of contact of the trophoblast with the uterus, a feature common to all mammals is that the trophoblast epithelium produces progesterone at some stage of gestation. High tissue concentrations of progesterone have been shown to suppress maternal cellular response, and this may be one important mechanism for preventing rejection. It may also explain why embryos will develop in ectopic sites, sometimes to full term. However, the two species of marsupial so far studied produce very little placental progesterone although they too have developed a tolerance of the conceptus, since presensitization of the mother to paternal antigens has no influence on gestation length. The marsupial mammals are like the eutherians in this regard, and it seems that both groups have developed a number of highly complex immunological interactions between the fetal–placental unit and the mother. But the general conclusion is that the uterus is basically a hostile rejection site, which must be modified by hormones to accept a fetus, whereas the embryo is not so fussy about where it grows.

Placental structure and function

Formation of the fetal membranes

The fetal membranes are often large and voluminous, blocking our view of the embryo. They are essential auxillary structures necessary for embryonic growth, and are derived from three basic extra-embryonic germ layers (ecto-, meso- and endoderm) which support the growth, nutrition, respiration and excretion of the embryo throughout pregnancy. The fetal membranes are extremely variable in shape and size, and are probably responsible for confusing more students of embryology than any other tissue!

The single-layered trophoblast (or trophectoderm as it is also commonly called) is the first of the cell layers to become an extra-embryonic membrane, and as we have seen, it has an important role in the attachment and implantation of the embryo. This layer later fuses with avascular mesodermal cells to become the chorion, the outer envelope that encloses the entire embryo and the other three fetal membranes, the amnion, yolk sac and allantois (Fig. 2.14).

The yolk sac occurs in all embryos with an amnion, and in the mammals it develops early from the blastocoele cavity. In reptiles and birds the yolk sac of the freshly ovulated egg encloses quite large quantities of yolk, but in the mammalian egg there are only yolk bodies or very small yolk masses; and even in the monotremes (platypus and echidna), which lay eggs, most of the accumulated nutrients are derived from the uterus. After fusion with the chorion, the yolk sac may be vascular (trilaminar) or avascular (bilaminar), and the chorio-vitelline placenta thus formed makes the first attachment, providing a site for exchange with the mother. In some mammals, such as the human, the yolk sac becomes completely vestigial after a week or so, whilst in others, such as the rabbit, the yolk sac placenta remains as an important organ of nutrient and antibody exchange throughout pregnancy.

The amnion develops from extra-embryonic ectoderm and avascular mesoderm so as to surround completely the embryo proper. It provides a fluid-filled environment in which the embryo can float and develop in a state of weightlessness. The amniotic fluid also provides protection from dessication and mechanical shock. The amnion is never vascularized, although it may have this appearance in some species if it is overlain by the vascular yolk sac.

The allantois is essentially an outgrowth of the embryonic hindgut, and is continuous with the urinary bladder. Allantoic fluid is therefore fetal urine, rich in urea and nitrogenous waste products. The allantois is derived from endoderm and vascular mesoderm, and as it expands into the extra-embryonic coelom it may fuse with the chorion to form the 'definitive' chorio-allantoic placenta. In most mammals the chorio-allantois takes over from the chorio-vitelline placenta as the main organ

Fig. 2.14. Pattern of development of the fetal membranes in a generalized mammal, based largely on the arrangement seen in the pig. In 5, 6 and 7, the stages shown are just before and during the elongation of the embryonic vesicle, which occurs in the long dimension of the uterus. (From J. S. Perry. *J. Reprod. Fert.* **62**, 321–35 (1981).)

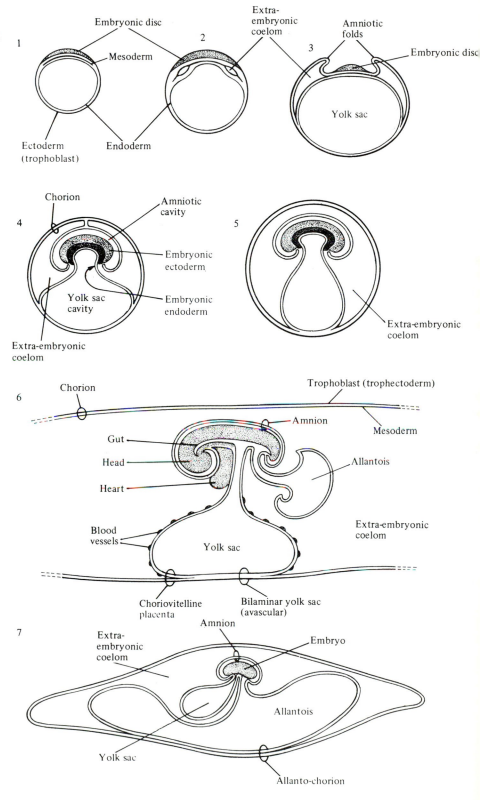

of respiratory and nutritive exchange, but in many marsupials the allantois remains enfolded by the wall of the yolk sac and never reaches the chorion.

At this point it is perhaps necessary to emphasize that there is great variety in placental structures seen among mammals, but placental morphology gives no clue to phylogeny. The most 'primitive' placentae are not necessarily possessed by the most 'primitive' mammals. It is also incorrect to refer to eutherian mammals as 'placental' mammals, for as we shall see, marsupials also have perfectly respectable placentae!

The monotremes represent a rather special case. One of these mammals, the platypus, when first discovered and sent back to Europe, was considered as a rather bad scientific joke, for the stuffed specimen was equipped with a soft leathery bill like a duck, dense, waterproof fur like an otter, webbed feet, no nipples or teats despite the fact that it was reputed to secrete milk, and to crown it all, it was claimed to lay eggs! It took nearly 100 years to convince the scientists on this latter point, although as Mervyn Griffiths points out, many people living in Australia in the nineteenth century knew they were oviparous; the final evidence was provided by a young Scot, W. H. Caldwell, working in Australia, who sent the now famous telegram to the British Association meeting in Montreal in 1884: 'Monotremes oviparous, ovum meroblastic'.

Although unquestionably egg-layers, the term 'oviparous' does not strictly describe the echidna and platypus. Cleavage is meroblastic (that is, confined initially to the top surface of the egg) because, as in the hen's egg, the yolk does not participate in the cleavage. However, the quantity of yolk in a monotreme egg is not sufficient to provide all the nutrients for the developing conceptus; at the time of ovulation, the egg is only 4.4 mm in diameter, and thus considerably smaller than those of comparably sized birds and reptiles. Bill Luckett has recently studied the monotreme collections made at the end of the nineteenth century and now housed in the Hubrecht Laboratories in Holland; he describes how the trophoblast (epiblast) rapidly divides and surrounds the yolk mass to form a 'bilaminar' blastocyst, which then undergoes a rapid distension due to the absorption of uterine secretions, eventually reaching a diameter of 12 mm. This absorption of uterine fluids occurs through the two-layered shell which was laid down in the oviduct and in the uterus, and which surrounds the trophoblast. The third layer of the shell, secreted by the uterine glands, becomes apparent after the blastocyst is 12 mm in diameter. Leon Hughes in Brisbane has shown that the shell is porous, which would allow relatively easy passage of nutrients and oxygen.

Intrauterine growth of the egg is unknown in oviparous reptiles and birds, and Bill Luckett suggests that this development of the monotreme trophoblast for nutritive exchange perhaps foreshadows its functions in the eutherians and marsupials. The egg is laid when the embryo has formed 18 somites in the platypus and 19 somites in the echidna, but the familiar arrangement of the fetal membranes is not seen until about the 30–35-somite stage (Fig. 2.15). The allantois does not contact the chorion until the 40-somite stage, so that until this time the yolk sac is the only site of nutritive and respiratory exchange; subsequently both membranes appear to function in respiratory exchange.

Fig. 2.15. (*a*) Platypus embryo of 35 somites from incubated egg. The amnion is closed by an extensive chorio-amniotic connection. A sinus terminalis marks the junction between the chorio-vitelline membrane (= trilaminar yolk sac) and the persisting bilaminar yolk sac. (*b*) Definitive arrangement of the fetal membranes of the platypus in a late incubation stage (which occurs in the pouch-like area near the mammary glands after laying). Note the extensive area of both chorio-allantoic and chorio-vitelline membranes, and the persisting chorio-amniotic connection. (From W. P. Luckett. In *Major Patterns in Vertebrate Evolution* ed. M. K. Hecht, P. C. Goody and B. M. Hecht. Plenum Press; New York. (1977).)

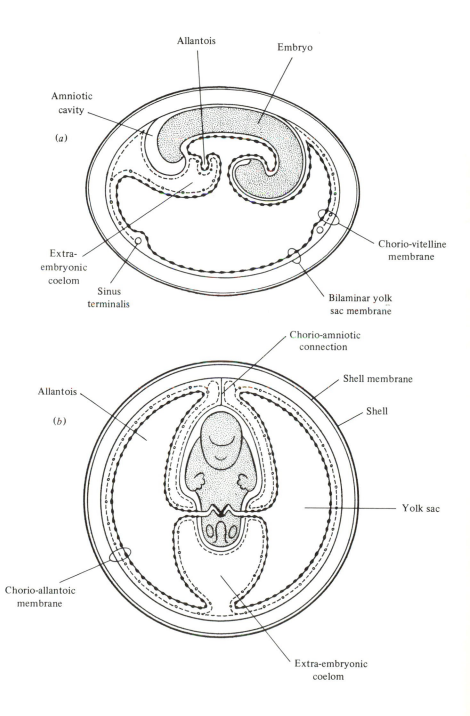

Placental types

The classification of placental types has been an obsession of numerous scientists over the past 100 years, working at a time when anatomists in general fervently believed in the dogma 'ontogeny recapitulates phylogeny'. One concept of placental classification, first put forward by Grosser in 1909, was that the efficiency of the placenta is inversely proportional to the thickness of the barrier (or number of cellular layers) separating the maternal and fetal circulations. Thus placentae were classified as *epitheliochorial* (three maternal layers and three fetal layers) *syndesmochorial* (two maternal and three fetal), *endotheliochorial* (one maternal, three fetal) and *haemochorial* (no maternal layers; free maternal blood bathes three fetal layers). Harland Mossman in 1926 added another type, *haemoendothelial* (free maternal blood bathing only one fetal layer, the vascular endothelium) (Fig. 2.16).

Fig. 2.16. There are six cellular layers that are potential barriers to the transport of gases, nutrients and other substances between the blood of mother and fetus. The Grosser classification depended on the number of layers in each type of placenta. A = epitheliochorial (6 layers, e.g. mare, pig, cow, sheep); B = endotheliochorial (4 layers, e.g. seal, dog, cat, ferret; C = haemochorial (3 layers, e.g. man, rabbit, rat, mouse).

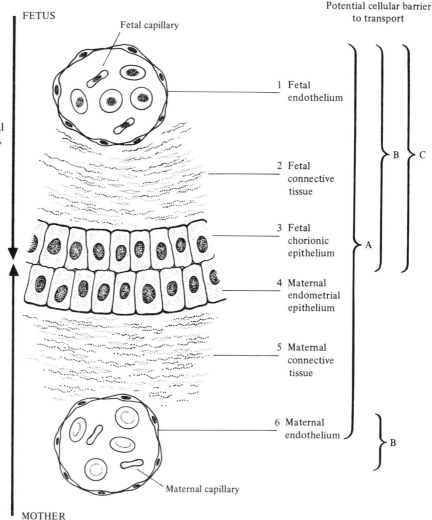

This classification implied that diffusion or transport across a multi-layered placenta, such as an epitheliochorial one, must by definition be inefficient, whilst the haemochorial placenta was the most efficient. The classification has been useful for categorizing different placental types, but its functional implications have been disproved, thanks in particular to the work of Professor E. C. Amoroso – Amo to his friends.

For example, in eutherian mammals it is often the case that the greater the number of placental layers, the more fully developed the neonate is at birth; compare the newborn foal and the newborn human infant. Donald Steven in Cambridge has summarized the controversies surrounding placental classification, and provided a simpler version which does not attempt to describe the finer details of the placental barrier. He lists three main placental types amongst the mammals, but even with such groupings he emphasizes that variations in structure can occur even within the same category (Table 2.2).

There are other useful systems of placental classification, based on gross morphology. In 1604 Fabricius introduced a classification based on the appearance of the sites of chorionic attachment to the endometrium. He listed four main placental types, now known as diffuse, cotyledonary, zonary and discoidal (Fig. 2.17).

Fig. 2.17. Illustration of the four main types of mammalian placenta: diffuse, cotyledonary, zonary and discoid. Diffuse placentae, where there is an attachment of the chorionic sac over its entire surface, are found in the horse, pig, camel, many dolphins and whales, and in the lemurs. Cotyledonary placentae, in which the chorion has specialized villi restricted to circular or oval areas over the chorionic sac, are characteristic of the ungulates. Zonary placentae may be complete as shown, in which the chorionic villi are restricted to an equatorial girdle, bounded by the (green) haemophagous organ, as found in the dog, cat and genet, or incomplete (circular or equatorial) regions as in ferrets, racoons, mink and bears. The haemophagous organ in these types may be central or marginal, distinguishing them from the discoid type which has no marginal effusion of blood. Finally, the discoid placenta characteristic of man, rat, mouse or guinea pig is single, and the chorionic villi are arranged in a circular plate, or may be double as in macaque monkeys.

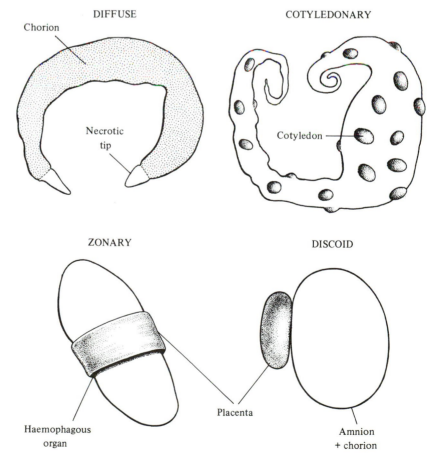

DIFFUSE

Chorion

Necrotic tip

COTYLEDONARY

Cotyledon

ZONARY

Haemophagous organ

DISCOID

Placenta

Amnion + chorion

The diffuse placenta

The diffuse placenta is found in a wide range of species, including pigs, horses, camels, lemurs, moles, whales, dolphins, kangaroos and possums. The villi of the chorion are distributed more or less evenly over the entire surface of the chorionic sac. The villi interdigitate with corresponding depressions or villi in the uterine epithelium, and physiological exchange takes place across these surfaces.

The most striking feature of the ontogeny of the fetal membranes in the pig is that the membranes undergo a period of rapid and dramatic elongation between days 6 and 12 of gestation, during which time the 2 mm spherical vesicle grows into a filament of up to 1 m in length (see Fig. 1.2). This elongation is due to a proliferation of trophoblastic tissue. The allantois first appears at day 14 and by day 26 the chorion is vascularized by the allantoic blood vessels, although the allantois does not penetrate to the tips of the chorionic sac, which have a necrotic appearance.

Although it is well recognized that the fetal placenta can produce gonadotrophic hormones during pregnany in many species (see Book 3, Chapters 1 and 3), the horse always appeared to be an exception to the general rule; equine chorionic gonadotrophin was known to be produced by ulcer-like structures on the inner surface of the uterus, the endometrial cups, and hence it was assumed to be a maternal hormone. The studies of Twink Allen and his colleagues in Cambridge have now shown conclusively that the endometrial cup tissue is fetal in origin after all, and its mode of formation and regression are of great fundamental interest.

Table 2.2. *Mammalian placental classification.* (After Steven (1975) – see Suggested Further Reading.)

Histological type	Mammalian order	Common name
Epitheliochorial	Perissodactyla	Horses, rhinos, tapirs
	Cetacea	Whales, dolphins, porpoises
	Lemuridae	Madagascan lemurs
	Artiodactyla	Cloven-hoofed ungulates, cattle, sheep, pigs, deer, antelopes, giraffe
Endotheliochorial	Most Carnivora	Felids, mustelids, bears, dogs
	Pinnipedia	Seals
Haemochorial	Insectivora	Shrews
	Rodentia	Rats, mice, etc.
	Lagomorpha	Rabbits, hares
	Sirenia	Dugong, manatee
	Most Primates	New and Old World monkeys, apes, man

The horse blastocyst expands to form a large, spherical vesicle before implantation (Fig. 2.18). By day 21 the amnion is fully formed and surrounds the embryo, and the yolk sac is very large. The border of the vascular (bilaminar) yolk sac is marked by the sinus terminalis, with an area of bilaminar, avascular yolk sac persisting beyond it. The allantois is at this stage still small and has not yet contacted the chorion. By day 35 the allantois has enlarged and fused with the chorion at the embryonic pole, thereby surrounding the amnion (Fig. 2.19). The vascular yolk sac has now

Fig. 2.18. The growth and expansion of the embryonic vesicles in (*a*) pony mares, and in (*b*) tammar wallabies up to the time of attachment. There is a close similarity between the two species at this stage of development ((*a*) From K. J. Betteridge. In *Embryo Transfer in Farm Animals*, ed. K. J. Betteridge. Canada Department of Agriculture Monograph 16 (1977). (*b*) From M. B. Renfree, unpublished).)

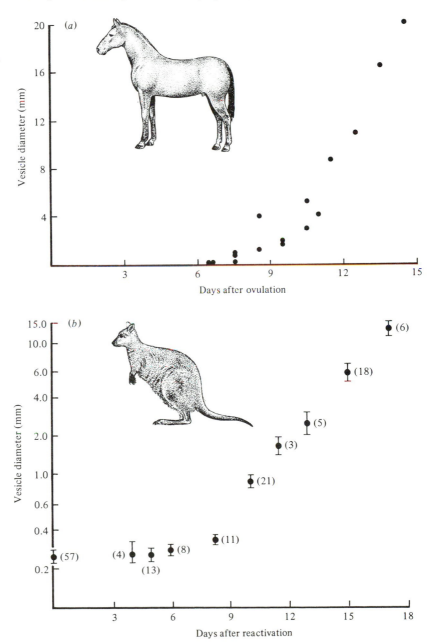

Fig. 2.19. Embryos of the
horse at day 25 (*a*) and day
35 (*b*) after conception, and
diagrams of the fetal membranes at these two stages.
A photograph of the
membranes of late pregnancy
is shown in (*c*) and an
explanatory diagram (both
on the facing page). The
white rectangle is a 6 inch
ruler ((*a*) and (*b*) kindly
provided by Dr W. R.
Allen, (*c*) by Professor
R. V. Short.)

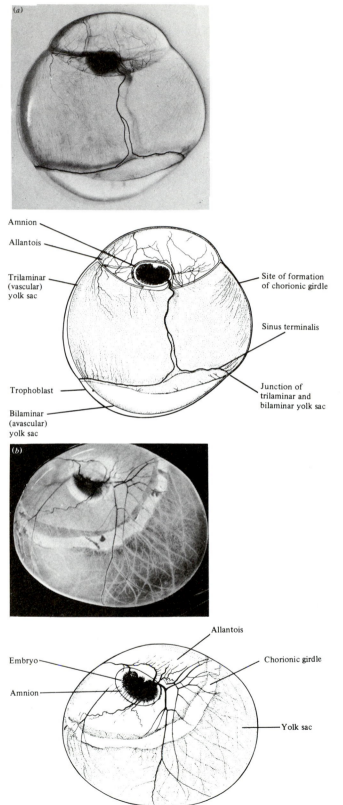

extended towards the abembryonic pole, further reducing the avascular yolk sac. The first attachment to the endometrium occurs at this stage. A distinct belt of elongated trophoblast cells forms around the circumference of the chorionic sac just below the margins of the allantois; this structure is known as the chorionic girdle. By 42 days of gestation, the allantois has fused with almost the entire chorion, and the chorionic girdle remains below the allantois, now close to the abembryonic pole. Starting on about the 35th day of gestation, cells become detached from the girdle to penetrate the maternal endometrium and burrow deep into the stroma,

Fig. 2.19 (*cont.*)

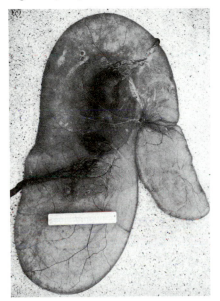

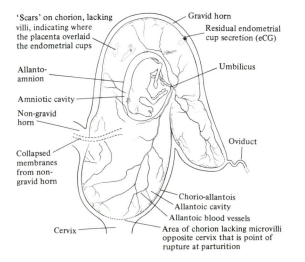

'Scars' on chorion, lacking villi, indicating where the placenta overlaid the endometrial cups

Gravid horn

Residual endometrial cup secretion (eCG)

Allanto-amnion

Umbilicus

Amniotic cavity

Non-gravid horn

Oviduct

Collapsed membranes from non-gravid horn

Cervix

Chorio-allantois
Allantoic cavity
Allantoic blood vessels
Area of chorion lacking microvilli opposite cervix that is point of rupture at parturition

where they enlarge to form the characteristic 'decidual cells' of the endometrial cup. On about the 40th day, the cup tissue becomes visible to the naked eye as a band running around the circumference of the pregnant uterine horn, and equine chorionic gonadotrophin first appears in the maternal circulation. Endometrial cup development and gonadotrophin levels increase to maximum by about day 60, and thereafter the gonadotrophin titres begin to fall as the decidual cells become surrounded by a mass of lymphocytes. Eventually, the cup tissue is sloughed off from the surface of the uterus into the uterine lumen, a process that bears a remarkable histological resemblance to a typical graft rejection reaction.

If a mare is covered by a jack donkey to produce a mule fetus, the hybrid endometrial cups stimulate an enhanced lymphocytic response, and are sloughed off prematurely, with the result that equine chorionic gonadotrophin is barely detectable in the maternal circulation, although the pregnancy is normal in all other respects. If a jenny donkey is covered by a stallion, to produce a hinny fetus, the endometrial cups are excessively developed and the concentrations of equine chorionic gonadotrophin are extremely high. Thus it seems that gonadotrophin production is determined by the genotype of the fetus; the mother apparently becomes 'aware' of the fetal cells that have invaded her endometrial stroma and mounts a classical immunological response to reject them. As we have seen, fetal tissue is normally protected from maternal immunological attack, so the endometrial cups of the horse provide an intriguing exception to the rule.

One question that remains unanswered is why the horse needs to produce such a large quantity of gonadotrophin. There may be over a million international units of equine chorionic gonadotrophin present in the circulation of the mare at any one time, and yet this massive amount probably does not cause any follicular development in the maternal ovaries. However, it does result in the formation of a number of accessory corpora lutea either by ovulation or by luteinization of spontaneously developing follicles, and they may act as a supplementary source of progesterone in early gestation. If a mare aborts after the 35th day of gestation, when the fetal cells have already invaded the endometrium, the cups continue to develop normally and secrete equine chorionic gonadotrophin and accessory corpora lutea may be produced. Thus the mare fails to return to oestrus, and has lapsed from pregnancy into pseudopregnancy. It is ironical that equine chorionic gonadotrophin, which is so potent at stimulating follicular development in so many species, from mice to women, appears to be completely inactive in this regard in the Equidae, where it only seems to exert weak, luteinizing hormone-like activity.

The subsequent development of the horse fetal membranes is by expansion of the chorionic sac, which extends into both the gravid and non-gravid horns of the uterus, with the chorio-allantoic placenta being in contact with the endometrium. The complete outer surface of the chorion, with the exception of the area overlying the sites of endometrial cups and the cervix, is covered in tufts of short villi – microcotyledons, The yolk sac is reduced to a vestigial structure within the allantoic part of the umbilical cord, although still attached to the chorion.

The marsupials do not fit conveniently into the placental classification types proposed for the eutherian mammals. Most marsupials have a placenta that most closely resembles the 'diffuse' type, although the attachment is through the yolk sac and not the allantois. Until recently

Fig. 2.20. Embryos of the tammar wallaby at: (*a, b*), day 17 (diameter of vesicle, 15 mm); (*c*) day 20; (*d*) day 21 (crown–rump length 8 mm) and (*e*) day 25½ (crown–rump length 16 mm) after reactivation of the diapause blastocyst. There is some similarity in the early stages to the horse blastocyst (see Fig. 2.19), but the allantois never makes contact with the chorion to form a placenta. Birth occurs on day 27.

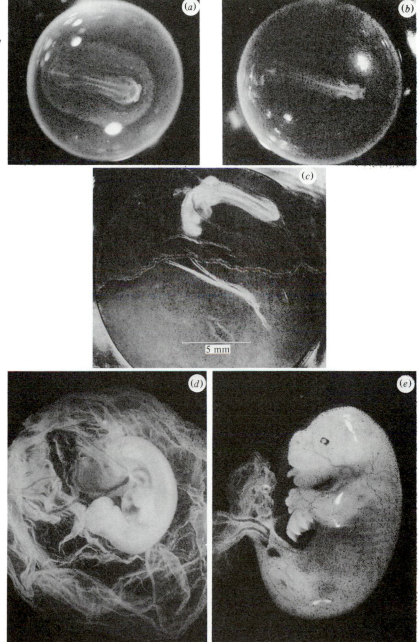

no study had been made of placental exchange or hormone production in marsupial placentae, so it was assumed that they were inefficient and incapable of steroid synthesis; we now know this assumption to be incorrect. Marsupials may live 'down-under', but they shouldn't be under-rated!

In the wallaby, the early stages of blastocyst expansion leading to placental formation are similar in rate of enlargement to those in the horse (Fig. 2.18). As already mentioned, marsupial blastocysts are unilaminar spheres lacking an inner cell mass, so that it is not possible to determine which of the ectodermal cells will form trophoblast and which the embryo proper. Early development occurs primarily by expansion, and a bilaminar blastocyst is formed by 9–10 days. Mesoderm then appears between the ectoderm and endoderm, and grows out to form the vascular yolk sac, delineated by the sinus terminalis (Fig. 2.20). By 19 days the amnion is fully formed around the embryo, and the yolk sac placenta becomes attached to the endometrium by about day 20. The allantois appears at this time as a tiny sac, but unlike its counterpart in the horse, it remains enfolded in the wall of the yolk sac and never reaches the chorion to form a chorio-allantoic placenta (Fig. 2.21). However, some marsupials, such as the bandicoots, *do* develop chorio-allantoic placentae, which may be highly invasive – another reason for not restricting the term 'placental' mammals to eutherians. No study has yet been made of the endocrinology of the marsupial chorio-allantoic placenta, but Helen Padykula and Mary Taylor have shown that ultrastructurally there is a very intimate contact between maternal and fetal tissue. In the wallaby, the bilaminar, avascular yolk sac persists throughout pregnancy, although its surface area is reduced to about one-third of the total by full-term at 26–27 days. Early

Fig. 2.21 Diagram of the fetal membranes of the tammar wallaby. Note that the allantois cannot make contact with the chorion.

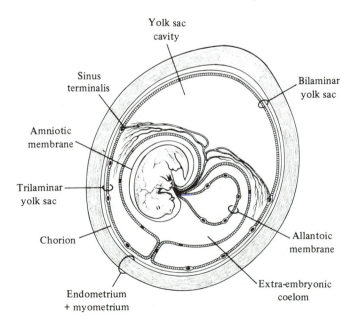

workers suggested that nutrient exchange took place across the bilaminar yolk sac, whilst respiratory exchange was the function of the trilaminar yolk sac, becoming most important close to term; recent biochemical and physiological measurements support this idea. At parturition, the wallaby placental membranes surround the fetus and there is no evidence that it is retained and resorbed in the uterus – a type of placenta found in moles and classified as 'contradeciduate'. This type was previously said to occur in the marsupials.

The cotyledonary placenta

The cotyledonary placenta is characteristic of the ruminants; instead of being uniformly distributed over the entire surface of the chorion, the chorionic villi are clumped together into well developed circular regions known as cotyledons. These cotyledons develop only in those regions of the chorion that overlie predetermined aglandular areas of the endometrium known as the caruncles. The fetal cotyledon and the maternal caruncle unite to form a placentome, and these placentomes are the only sites of maternal–fetal exchange, the intercotyledonary chorion being devoid of villi and unattached to the endometrium. The number of caruncles varies greatly among species from as few as three or four per uterine horn in the roe deer, reindeer and Père David's deer up to 180 in the goat and giraffe. It is interesting to note that when Leonardo da Vinci drew his famous illustration of the human fetus *in utero*, he provided the baby with a cow's cotyledonary placenta and even drew in the fine detail of a placentome, showing the villous nature of the cotyledon and the caruncular pits with which it interdigitates (Fig. 2.22). It is interesting to speculate why he

Fig. 2.22. Leonardo da Vinci's drawing of the human fetus showing it attached to the uterus by a cow's cotyledonary placenta. (Reproduced by gracious permission of Her Majesty the Queen.)

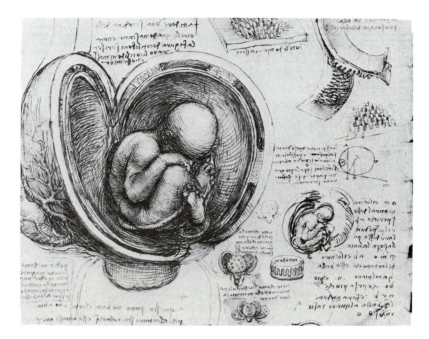

made this mistake, and his act serves as a warning to those who make unwarranted interspecific comparisons.

The ruminant conceptus elongates before forming a placental attachment, but not to such a dramatic extent as in the pig. In the sheep the embryo grows from a 1 mm sphere to a 1-m elongated sac between the twelfth and fourteenth days of gestation. By day 15, small areas of trophoblast have started to attach to the caruncular epithelium, but it is not until about day 30 that placentome formation is obvious to the naked eye; the final arrangement of the fetal membranes and placenta is shown in Fig. 2.23. Similar changes are seen in other ruminants. In the roe deer, the only ruminant with delayed implantation, the rapid elongation takes place after the termination of embryonic diapause, and the formation of the

Fig. 2.23. Arrangement of the fetal membranes of ungulates. (*a*) A cow embryo, crown–rump length 95 mm, 70 days after mating, within its allanto-chorionic sac. Cotyledons can be seen on the surface. (*b*) Progressive development of the fetal membranes in the sheep. (From H. W. Mossman. *Contributions to Embryology*, No. 158. American Anatomical Society (1937).)

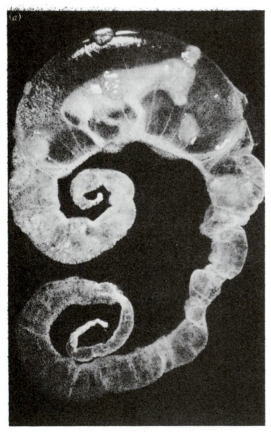

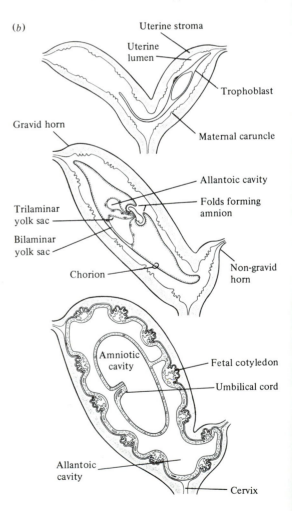

placentomes (usually only four per embryo) follows in a fashion similar to that seen in the sheep.

The zonary placenta

The zonary placenta is characteristic of the carnivores, and is the result of an aggregation of chorionic villi to form a broad band that encircles the equatorial region of the chorionic sac. Zonary placentae may be

Fig. 2.24. Early (*a*) and late (*b*) stages of development of the incomplete zonary placenta of the African elephant. Trophoblastic villi (t) can be clearly seen in the zonary attachment band in the early stage, but are absent outside the band. The later stage shows the chorioallantoic sac, with one of the three interruptions in the placental band. (From E. C. Amoroso and J. S. Perry. *Phil. Trans. Roy. Soc.* **B248**, 1–34 (1964).)

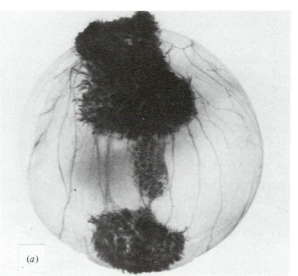

(*a*)

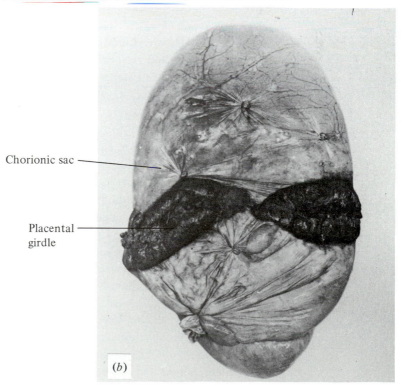

Chorionic sac

Placental girdle

(*b*)

complete, as in dogs and cats, or incomplete, as in bears, seals, and mustelids like the ferret and mink. In the dog and cat, the yolk sac membrane forms a functional chorio-vitelline placenta early in gestation whilst the chorio-allantois is developing equatorially. The chorio-allantois erodes the uterine epithelium, so that the cells of the cytotrophoblast and their enclosed allantoic mesoderm come into close relationship with the maternal blood capillaries. The yolk sac persists as a vestigial structure floating in the allantoic fluid, whilst the chorio-allantois remains as an oblong, fluid-filled sac, with its girdle of placental villi. Incomplete zonary placentae may resemble the single or double discoid type (see next section), but the zonary placenta always has a central or marginal effusion of maternal blood (the haemophagous organ). In dogs, this forms a bright green margin to the vivid red of the placental attachment zone.

Zonary placentae are also present in the elephant, and related species such as the dugong, manatee and hyrax. They seem to develop from a diffuse placenta, in which the chorionic villi at the polar areas of the chorionic sac subsequently regress. In the African elephant, villi are only found in the equatorial region of the blastocyst and are never diffuse. The placenta sometime extends completely around the equatorial zone, but is more usually interrupted at one or more places (Fig. 2.24). The chorion is completely separated from the amnion by the voluminous allantois, and the yolk sac apparently becomes vestigial between 2 and 3 months of gestation.

The discoid placenta

The discoid placenta is found in a mixed group of mammals, including man and mouse, bats and rats, rabbits, hares and insectivores – try to discern a taxonomic pattern in that group! The chorionic disc may be single (man) or double (monkeys of the Orders Cercopithecidae and Cebidae). We should remember, however, that not all primates have interstitial implantation resulting in the formation of a discoid placenta; strepsirhine monkeys (lemurs, galagos, etc.), for example, have a non-invasive, central mode of implantation (Fig. 2.25). In species with a discoid chorio-allantoic placenta, the yolk sac membrane nevertheless plays a brief role as the organ of nutrient exchange. In man, the yolk sac appears very early, around the second week, although it never forms a chorio-vitelline placenta; it has become constricted by the fourth week, and has regressed by the eighth week, when it only remains as a vestige in the umbilical stalk. In rodents and lagomorphs the yolk sac is important not only in early gestation as the site of nutrient exchange before the chorio-allantois has formed, but it remains functional later in pregnancy. After the formation of the chorio-allantois, the surface of the bilaminar yolk sac adherent to the endometrium breaks down, resulting in apposition of the *inner* surface of the trilaminar yolk sac membrane to the uterus. Although most exchanges take place across the chorio-allantois, this *inverted* yolk sac placenta remains an important route for the transmission of antibodies from mother to fetus (Fig. 2.26).

The formation of the fetal membranes and the process of implantation is probably best known for man and mouse, and in the later stages the placentae of the two species are relatively similar, except that the mouse has a more extensive decidual reaction, and its yolk sac is proportionally larger, becoming inverted in a similar manner to that of the rabbit (Fig. 2.26). In addition the mouse forms an 'egg cylinder' during the early stages of implantation, resulting in slight differences in the early formation of the fetal membranes (see Fig. 6.3).

In man, implantation is interstitial, and between days 7 and 12 the blastocyst penetrates the uterine epithelium and comes to lie completely embedded in the uterine stroma (Fig. 2.27). The trophoblast actively phagocytoses adjacent maternal tissues, maternal blood corpuscles, and

Fig. 2.25. Diagram showing the variation in types of implantation within the Order Primates (see also Fig. 2.3). (a) Non-invasive, central implantation of strepsirhine primates (lemurs, galagos, etc.). (b) Invasive, eccentric implantation in *Tarsius*. (c) Invasive, eccentric implantation of ceboids and cercopithecids (capuchin monkeys, rhesus and other macaques, etc.). (d) Interstitial implantation of hominoids (great apes and man). (From W. P. Luckett. In *Phylogeny of the Primates*, ed. W. P. Luckett and F. S. Szalay, pp. 257–82. Plenum Press; New York (1975).)

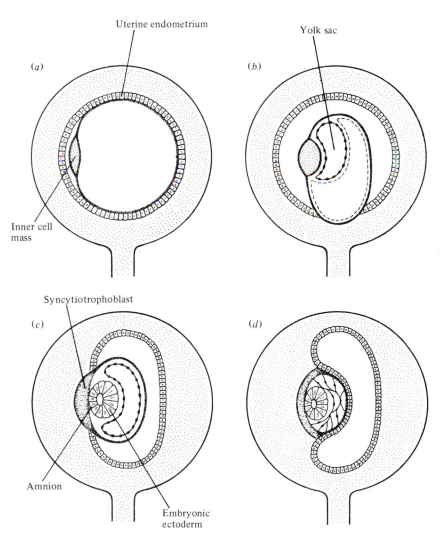

the secretions of the uterine glands. The trophoblast differentiates into an irregular *cyto*trophoblast surrounded at the point of invasion by an extensive outer syncytium, the *syncytio*trophoblast. The amnion exists as a closed cavity from its earliest appearance, derived from a space between

Fig. 2.26. Diagrams of the arrangement of the fetal membranes of the rabbit and mouse. Note that both have an inverted yolk sac placenta (in which the outer surface has broken down, leaving only the inner, or endodermal surface apposed to the uterus) as well as the discoid allantoic placenta.

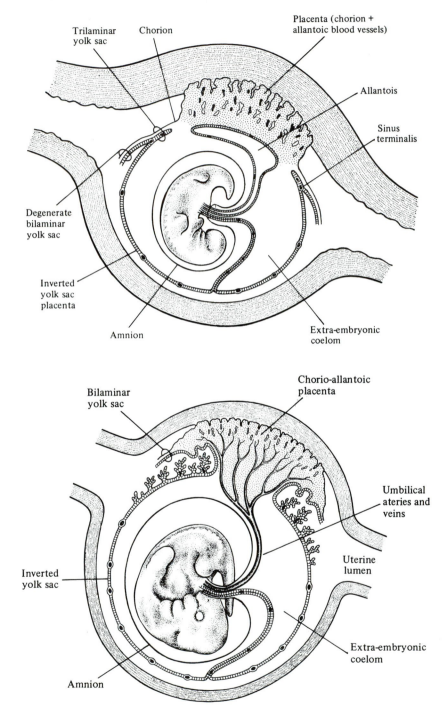

the ectodermal cells of the embryo and the region of trophoblast immediately above. The allantois, like that of the mouse and other rodents, grows out from the terminal part of the hindgut towards the yolk sac stalk. The allantois grows down the umbilical stalk along with the umbilical blood vessels, but it never becomes fluid-filled, and after extending along the entire length of the umbilical cord, it degenerates distally (Fig. 2.28). The placenta, which is clearly defined by the third month of pregnancy, thus consists of the chorion vascularized by allantoic mesoderm. By the fourth month the placenta consists of three layers – the trophoblast, a thin sheet of connective tissues, and the fetal vascular endothelium (see Fig. 2.16). Exchange occurs both by passive diffusion, and more importantly, by active transport.

Decidualization of the uterine mucosa results in three types of decidua – the *decidua basalis* (underneath the site of original invasion), the *decidua capsularis* (surrounding the conceptus and separating it from the uterine lumen) and the *decidua parietalis* (on the uterine wall opposite the implantation site). From the fourth month onwards, the decidua parietalis and capsularis come in contact, fuse, and obliterate the uterine cavity (Fig. 2.28). The final arrangement of the fetal membranes therefore consists of the amnion, closely applied to the chorion, which is in turn so closely applied to the fused deciduae that they appear as one membrane enclosing

Fig. 2.27. Three-dimensional diagram of a human blastocyst that has penetrated into the uterine mucosa, showing the direction of proliferation of the trophoblast. Lacunae in the syncytiotrophoblast are lakes of maternal blood. (From H. Tuchmann-Duplessis, G. David and P. Haegel. *Illustrated Human Embryology*, vol. 1. Springer-Verlag, New York; Chapman and Hall, London; Masson et Cie, Paris (1971).)

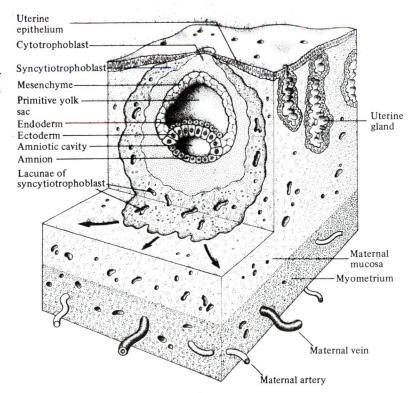

Uterine epithelium
Cytotrophoblast
Syncytiotrophoblast
Mesenchyme
Primitive yolk sac
Endoderm
Ectoderm
Amniotic cavity
Amnion
Lacunae of syncytiotrophoblast
Uterine gland
Maternal mucosa
Myometrium
Maternal vein
Maternal artery

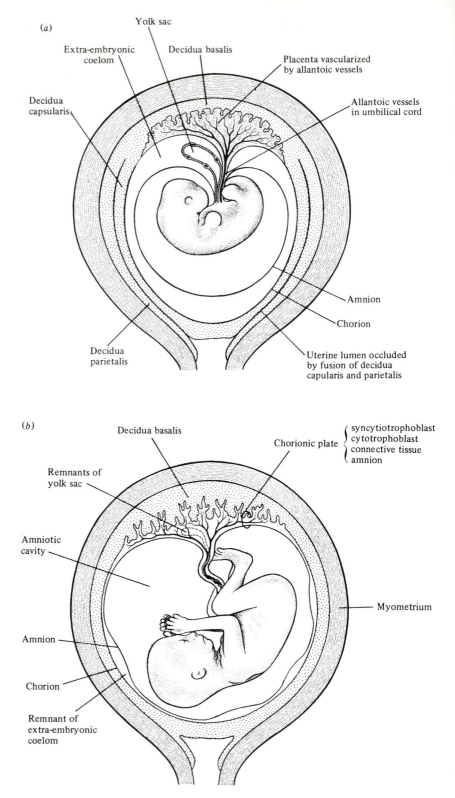

Fig. 2.28. Arrangement of the fetal membranes of the human embryo at around 40 days (a) and by the fourth month (b). Note the loss of the yolk sac, and the fusion of the amnion and chorion in the later stages.

(a)

Yolk sac

Extra-embryonic coelom

Decidua basalis

Placenta vascularized by allantoic vessels

Decidua capsularis

Allantoic vessels in umbilical cord

Decidua parietalis

Amnion

Chorion

Uterine lumen occluded by fusion of decidua capularis and parietalis

(b)

Decidua basalis

Chorionic plate

{ syncytiotrophoblast
cytotrophoblast
connective tissue
amnion

Remnants of yolk sac

Amniotic cavity

Myometrium

Amnion

Chorion

Remnant of extra-embryonic coelom

Fig. 2.29. Arrangement of the fetal membranes in different types of monozygotic twin pregnancies. *Type 1:* separate membranes = dichorionic twins. Separation of the zygote occurred at the two-blastomere stage (1–3 days after fertilization). *Type 2:* the most frequent: one placenta, separate amnions = monochorionic diamniotic twins. Separation of the zygote occurred at the inner cell mass–early blastocyst stage (3–8 days) *Type 3:* completely common membranes = monoamniotic twins. Separation at the embryonic disc stage (8–12 days). *Type 4:* completely common membranes and yolk sac, and only partial separation of embryos, which can take place at any of several later stages = Siamese or conjoined twins. Type 1 occurs in dizygotic twins, but Types 2, 3 and 4 can only involve monozygotic twins. (Partly redrawn from H. Tuchmann-Duplessis, G. David and P. Haegel. *Illustrated Human Embryology*, vol. 1. Springer-Verlag, New York; Chapman and Hall, London; Masson et Cie, Paris (1971).)

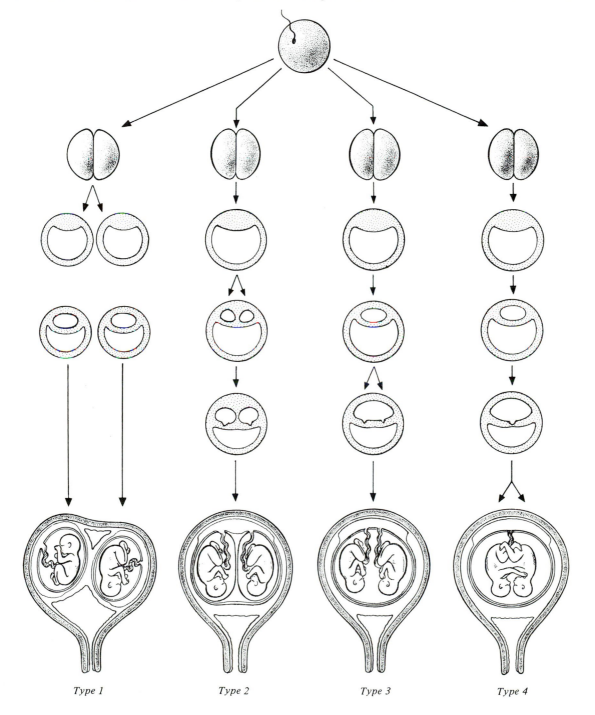

Type 1   *Type 2*   *Type 3*   *Type 4*

the amniotic cavity; the decidua basalis plus syncytiotrophoblast form the placenta proper, attached to the embryo by the umbilical stalk.

In the case of monozygotic twin pregnancies, the arrangement of the fetal membranes can give accurate information of the time of cleavage of the single zygote to form two individuals (Fig. 2.29). If the twins have completely separate chorionic sacs, cleavage must have occurred 1–3 days after fertilization, whereas at the other extreme, when they have a common amnion, cleavage must have been delayed until 8–12 days after fertilization. Conjoined ('Siamese') twins result if cleavage has been delayed until an even later stage. Theologians have used this evidence to determine when 'ensoulment' occurs. Assuming that monozygotic twins have separate souls, it follows that 'ensoulment' must occur *after* zygote cleavage, or at least 12 days after conception! If you can accept this line of reasoning, it would mean that contraceptive methods that prevent implantation, like the 'morning after' pill or the intrauterine device, should be acceptable to the Catholic church, which believes that true human life begins with ensoulment.

The way in which the developing embryo first makes contact with its mother has always fascinated biologists, and in recent years the exploitation of those species with embryonic diapause has provided an ideal experimental tool for investigating many aspects of the physiology, biochemistry and endocrinology of implantation. Given the chance, the embryo will implant almost anywhere at any time. But in the narrow confines of a hostile uterus, implantation is brought under strict maternal control. Once firmly attached to the uterus, the placenta develops its full complexity in a bewildering variety of shapes and forms. Attempts to deduce phylogenetic relationships from placental morphology have so far been futile. All mammals have a trophoblast which functions as the main site of physiological exchange and this is even true of the egg-laying monotremes, whose eggs gain maternal nutrients during their brief sojourn in the uterus. There are many similarities between marsupial and eutherian placentae, and there is certainly no justification for distinguishing eutherians as 'placental mammals', which implies that marsupials do not have placentae. But why Nature has devised so many 'enigma variations' on the common theme of the placenta remains a mystery.

Suggested further reading

The mammalian fetal membranes. J. A. Perry *Journal of Reproduction and Fertility*, **62**, 321–35 (1981).

Some comparative aspects of implantation. W. A. Wimsatt. *Biology of Reproduction*, **12**, 1–40 (1975).

Cellular basis of interaction between trophoblast and uterus at implantation. S. Schlafke and A. C. Enders. *Biology of Reproduction*, **12**, 41–65 (1975).

Mechanisms of implantation of the blastocyst. A. C. Enders. In: *Biology of Reproduction: Basic and Clinical Studies*. Ed. J. T. Velardo and
B. A. Kasprow, III. American Congress of Anatomy. New Orleans (1972).
M. E. Clutter. Academic Press; New York (1978).
Embryonic diapause in mammals – a developmental strategy. M. B. Renfree. In *Mechanisms of Dormancy and Developmental Arrest*, pp. 1–46. Ed.
M. E. Clutter, Academic Press; New York (1978).
Comparative morphogenesis of fetal membranes and accessory structures.
H. W. Mossman. *Contributions to Embryology*, No. 158 (1937).
Placentation. E. C. Amoroso. In: *Marshall's Physiology of Reproduction*, 3rd.
edn. Ed. A. S. Parkes. Longmans; London (1952).
The Human Placenta. J. D. Boyd and W. J. Hamilton. Heffer; Cambridge (1970).
The Uterus. C. A. Finn and D. G. Porter. Elek Science; London (1975).
Embryonic Diapause in Mammals. Ed. A. P. F. Flint, M. B. Renfree and
B. J. Weir. *Journal of Reproduction and Fertility, Supplement*, **29** (1981).
Cellular and Molecular Aspects of Implantation. Ed. S. R. Glasser and
D. W. Bullock. Plenum Press; New York and London (1981).
Comparative Placentation. Ed. D. H. Steven. Academic Press; London, New York and San Francisco (1975).

3

Sex determination and differentiation

R. V. SHORT

Mammals, like other vertebrates, have abandoned the simple vegetative pattern of reproduction by which lower forms reproduce themselves, budding off of a genetically identical individual from the parent body (cloning), and in its place they have opted for the more complex sexual reproduction, which offers a number of advantages. The separation of pairs of diploid chromosomes at meiosis to give haploid gametes, and their fusion at fertilization to give a new diploid zygote, shuffles the genetic pack of cards for each new generation. This variability becomes infinite when we add the gene migration that can occur between haploid pairs of chromosomes at meiosis as a result of crossing-over, plus a pinch of spontaneous genetic mutation that can travel through the population like wildfire if sufficiently advantageous. Variability is the substrate on which natural selection acts, and mammals (together with fish, amphibians, birds and reptiles) probably owe their advanced state of development to their sexual mode of reproduction.

But sex creates problems, since it requires the development of extensive dimorphisms in anatomy, physiology and behaviour between the two sexes. The female releases relatively few large gametes from her gonads at infrequent intervals, and is specially adapted to carry and nurture the embryo within her reproductive tract, and to feed and care for the newborn young until they can lead an independent existence. The profligate male, on the other hand, produces millions of tiny gametes a day from his gonads, and his reproductive tract is designed only for their intromission into the female at a time of her choosing. Although he plays no role in the care of the embryo, he may have an important supportive role in protecting the newborn, and has to be anatomically and behaviourally adapted to compete with others of his sex for access to females. So the central question to be addressed in this chapter is how are all these sexual dimorphisms brought about?

The obvious solution would be to produce all sex differences by genetic control. If a species could develop a sexually dimorphic pair of sex chromosomes, we might imagine that the problem was solved. In mammals, for example, where the male has an XY sex chromosome constitution, and the female is XX, all we would need to do would be to concentrate all the autosomal genes for the myriad of male characteristics onto the male-specific Y chromosome. But this would in fact be an extremely difficult feat to achieve, since the sex chromosomes never pair with the autosomes at meiosis, so there is no chance of genetic interchange by crossing-over. There is even very little opportunity for chromosomal exchange between the X and the Y chromosome themselves, as they have an extremely short pairing segment. This means that the genetic make-up of the X and Y chromosomes must have remained relatively unchanged during the course of evolution.

So what alternatives are there? As usual, Nature has been incredibly ingenious, and seems to have opted for a very simple genetic control of gonadal sex; in some way, the presence of a Y chromosome in mammals results in the development of a testis, and in the absence of a Y chromosome, an ovary will develop. From this point onwards, the whole of the rest of sexual differentiation is delegated to hormones produced by the gonads. The subtlety of the system lies in the fact that these hormones can activate genes that can be located anywhere on any of the autosomes. Thus secondary sexual characteristics, as we call them, are hormonally mediated and normally confined to one or other sex (sex-limited), whilst gonadal sex determination is controlled by genes on a sex chromosome (sex-linked). Sexual differentiation should be viewed as a gradual sexual awakening of the somatic tissues that is hormonally mediated; it spreads from the testis to encompass first the adjacent reproductive tract by local hormonal diffusion, then by systemic spread to give a male imprint to the developing fetal brain. Subsequently, in pubertal and adult life testicular and ovarian hormones both play a part in the development of male and female secondary sexual characteristics, so that there is hardly a tissue or an organ of the adult that does not show some form of sexual dimorphism if we look closely enough.

The best way of discussing this fascinating but incredibly complex subject is to follow a logical sequence of events. We begin by considering Genetic Sex, and the nature and location of the genes that are involved in the initial genetic determining event. This naturally leads into Gonadal Sex, and the way in which these genes determine whether a testis or an ovary develops. Whether or not the gonads are capable of producing fertile gametes depends on the genetic make-up of the germ cells – Germ Cell Sex. We will then consider Hormonal Sex, describing the different hormones produced by the testis and ovary, and the development of the internal genitalia. We then move on to Brain Sex, to gain some understanding of how fetal testicular hormones can masculinize the brain by influencing its anatomical development. We can then discuss Phenotypic Sex, and how male and female sex hormones influence the development of different secondary sexual characteristics, particularly at puberty. In order to understand why these phenotypic characteristics develop in the first place, we must also consider Sexual Selection, and the evolutionary forces that have been at work to encourage the development of sexual dimorphisms. The sex hormones produced by the gonads then interact with the brain to determine Behavioural Sex. And we will conclude with a discussion of Legal Sex – a peculiarly human concept, since in law you *must* be either male or female. The titillating question of Illegal Sex we had best leave to Book 8 in the First Edition, where it is discussed fully!

Throughout these discussions, it will be repeatedly apparent that we have learned an enormous amount about the normal from studying the abnormal. Such a complicated sequence of events is liable to go wrong at almost any point along the line, and the anomalous individual that results provides us with a fascinating experiment of Nature that we would be hard-put to reproduce in the laboratory. Any individual showing some of the phenotypic characteristics of both sexes is vaguely defined as an intersex. We can be more specific in our definition if we know the nature of the gonad. When both ovarian and testicular tissue are present, either as separate entities, or combined in an ovotestis, we call that individual a 'true hermaphrodite' – a rather rare condition. Far commoner is the intersex with only one type of gonad, whom we refer to as a 'pseudo-hermaphrodite'; it is ascribed a sex depending on the appearance of the gonad, so a male pseudohermaphrodite would possess testes, although his external phenotype might be entirely female.

Armed with a classification and a terminology, let us begin to consider the intricacies of the subject.

Genetic sex

In mammals, the female is known as the homogametic sex because she has two X chromosomes; thus all the gametes that she produces are similar to one another in that they possess a single X chromosome. The male, on the other hand, is referred to as the heterogametic sex because he has an X and Y chromosome, and hence two distinct populations of spermatozoa, one X-bearing and one Y-bearing. Sex in mammals is therefore determined at fertilization, depending on whether an X- or Y-bearing spermatozoon fuses with the egg. Since the X chromosome is much larger than the Y chromosome, and contains a host of genes that are not present on the latter, many people have attempted to separate the two populations of spermatozoa so as to influence the primary sex ratio at conception. Indeed, this search has become the present-day equivalent to that of the ancient alchemists for the Philosopher's Stone which would transmute base metals into gold. In spite of many different approaches, including density gradient centrifugation, sedimentation, electrophoresis, selective antiserum absorption, pH change, and alterations in the time interval between insemination and ovulation, none of the many claims to success has ever been substantiated on subsequent investigation. The principle obstacles to success seem to be firstly that any difference in mass between the X and Y chromosomes is only a minute proportion of the total mass of the spermatozoon, which in any case shows considerable variability, and secondly that the genotype of the haploid gamete is not normally expressed in its phenotype – a topic that is discussed more fully in Book 1, Chapter 3.

Birds have developed a sex-determining mechanism that is the opposite of that seen in mammals. The female bird is the heterogametic sex, and by convention is designated ZW, whilst the male is homogametic, and ZZ. This ZW/ZZ nomenclature is also used in fish, amphibians and reptiles wherever it is the female that is heterogametic; some fish and amphibians however have evolved an XX/YY sex-determining mechanism. It is an interesting consequence of female heterogamety that sex is not determined at fertilization, since all the spermatozoa carry an identical sex chromosome complement, but rather at ovulation, when a Z- or W-bearing egg is shed from the ovary.

If mammals and birds can differ so radically in their sex-determining mechanisms, how did sex chromosome heteromorphism evolve in the first place? Mammals and birds are both thought to be derived from early reptilian stocks, so could present-day reptiles hold any clues? A peep into their sex lives reveals a truly amazing story. Most reptiles, such as the turtles, tortoises, crocodiles, alligators and lizards, have no heteromorphic pair of sex chromosomes. However, a different picture emerges in the snakes. The boas and pythons (Boidae) are thought to represent the ancestral snakes most closely, and they lack a morphologically distinguishable pair of sex chromosomes. The non-venomous temperate snakes, like the grass snake (Colubridae), although having an almost identical karyotype to the Boidae, do show some evidence of sex chromosome heteromorphism in the female. This has become extreme in the venomous

Fig. 3.1. Schematic representation of sex chromosomes in female snakes. The Boidae lack heteromorphic sex chromosomes. By contrast, the Colubridae show some heteromorphism in the fourth largest pair, and the Viperidae show distinct heteromorphism in this pair. The heteromorphic chromosomes are designated W and Z, since the female is heterogametic. (From J. J. Bull. *Quart. Rev. Biol.* **55**, 3–20 (1980).)

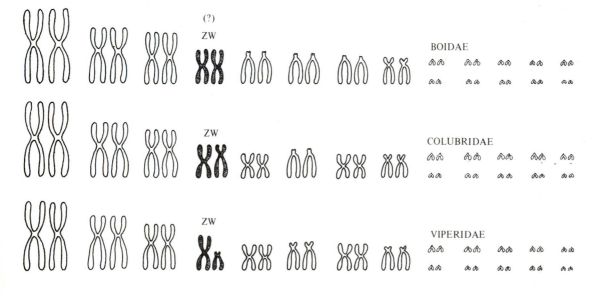

snakes, like the rattlesnakes and vipers (Viperidae), which are thought to be derived from the Colubridae and have clearly distinguishable Z and W chromosomes (Fig. 3.1).

Of course, just because we cannot *see* a heteromorphic pair of chromosomes does not necessarily mean that sex chromosomes do not exist. But this whole field took an exciting new turn when in the late 1960s and early 70s French scientists began to report the astonishing fact that the sex ratios at hatching in certain lizards and tortoises could be influenced by the temperature at which their eggs were incubated. The story was soon taken up by a number of workers in North America, and it has been abundantly confirmed. It now seems likely that all those reptiles that lack heteromorphic sex chromosomes probably lack a genetic mechanism for sex determination, and depend instead on incubation temperature to determine sex. The effects are quite dramatic, and a change of only 2–4 °C during incubation is all that is required. Turtle and tortoise eggs produce all males at incubation temperatures from 16–28 °C, and all females at temperatures above 32 °C, whereas lizards show the reverse picture, with all females at temperatures up to 26 °C, and all males beyond 28 °C. However, incubation temperature has absolutely no effect on the sex ratio in those lizards, snakes and turtles that have heteromorphic sex chromosomes. It is therefore attractive to suppose that the genetic control of sex, delegated to a specialized pair of chromosomes, evolved in our reptilian ancestors; certainly a temperature-dependent sex-determining mechanism would be of little use in a viviparous, homeothermic mammal, or even in a bird that goes to great lengths to incubate its eggs at a constant temperature.

Much work remains to be done on this temperature-dependent sex-determining mechanism in reptiles. We already know that there is only a brief critical period during embryonic development when temperature can influence sex, and it would be important to know whether that was during the period of gonadal differentiation, and if so, precisely how it operated. Even more fascinating are the ecological implications of this work. In the wild, the temperature of the eggs depends of course on the climate, and also on the site in which the eggs are laid. Jim Bull from Wisconsin has studied the nesting habits of the map turtle in the Mississippi river; it buries its eggs in sandbanks, and if the nest happens to be under a bush, the lower temperature results in a longer incubation period and a high proportion of males at hatching. Eggs laid in nests out in the open will be warmer, will therefore be the first to hatch, and will produce a high proportion of females. And consider for a moment the implications of this work for conservation. Most of the world's crocodiles, alligators and marine turtles are endangered species, and since they all lack heteromorphic sex chromosomes, it should be possible to increase the proportion of females for future breeding by incubating their eggs at the appropriate temperature.

But let us return to mammals. Having briefly considered how the sex chromosomes may have evolved, let us now discuss the nature and location of the genes that might be involved in sex determination. We can begin with the simple generalization that the Y chromosome is strongly male-determining. In its absence, the individual is almost invariably female, regardless of how many X chromosomes are present. But in its presence, the individual is almost invariably male, again regardless of how many X chromosomes are present. We will consider separately those fascinating but rare exceptions, the XX male and the XY female, which can be explained away without contradicting the above generalization.

We can make one more generalization about the Y chromosome: it contains very few genes. However, one gene that does seem to be common to the Y (or W) chromosomes of all mammals, birds, reptiles, amphibians and fish that have been examined is that which codes for a weak histocompatibility antigen (H) responsible for the rejection of tissue grafted onto an animal of the opposite sex. The antigen is therefore known as H-Y

Fig. 3.2. Genetical control of gonadal differentiation in mammals. H-Y antigen diverts the course of development so that the tissues of the indifferent genital ridge go to form testis; in the absence of the antigen they develop into ovary.

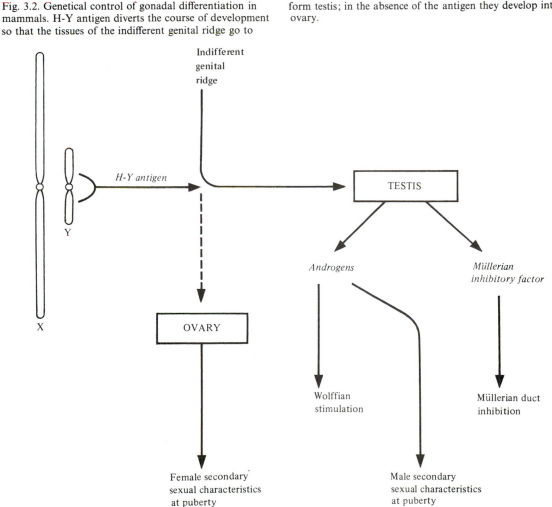

(or H-W in birds) and it can be detected in blood and tissues by a variety of immunological reactions. Its importance lies in the fact that the *H-Y* gene is the most promising candidate that we have for the male testis-determining gene (or *H-W* as an ovary-determining gene in birds); as a general rule the type of gonad that forms is in accordance with the individual's H-Y status (see Fig. 3.2). Most of our ideas about H-Y antigen we owe to the exciting work of Steve Wachtel in New York, and also of Susu Ohno in California whose chapter in Book 6 of the First Edition should be read in conjunction with this one.

From what we have said, you would expect that H-Y antigen could only be present if there was a Y chromosome, but this is not invariably the case. In mice, goats and humans one occasionally encounters H-Y antigen-positive XX males; in mice, we know that this condition is caused by an autosomal dominant gene called Sex Reversal (*Sxr*), and in goats it is caused by an autosomal recessive gene closely related to the dominant gene *Polled* (*P*), which also dictates whether the animals are harmless or grow horns. So the current view is that the *H-Y* gene exists in multiple copies, usually located around the centromere of the Y chromosome, but some copies of the gene can also be present on the X chromosome, or occasionally even on an autosome. In humans, some XX males show an extra piece of chromosomal material on one of the X chromosomes which has apparently been translocated from the Y.

Before we can really understand the male-determining role of H-Y antigen, we need to know a great deal more about this gene. Is it a structural gene, or a regulatory gene? It has been suggested for example that the true male-determining gene may actually be located on the X chromosome, and that H-Y antigen or autosomal genes merely regulate its activity. Does the indifferent gonad have to have receptors for H-Y antigen before the antigen can exert its effects? H-Y antigen is normally tightly bound to the cell surface, and the only cells that seem to be capable of secreting it are the Sertoli cells. But Susu Ohno has shown that some tumour cell lines can secrete H-Y antigen into their culture medium *in vitro*, and if the antigen is purified, and added to organ cultures of undifferentiated gonads from genetically female bovine fetuses, they will be transformed into testes. This experiment would therefore suggest that H-Y antigen can sometimes act like a circulating systemic hormone, to control gonadal development from afar. It would certainly provide an attractive explanation for the freemartin condition in cattle, in which a female fetus co-twin to a male, and sharing a common placental circulation (see Fig. 3.3), undergoes a partial sex-reversal of her gonads. Seminiferous tubules develop and apparently secrete Müllerian-inhibitory factor, whilst Leydig cells secrete testosterone, producing varying degrees of masculinization of the female's internal genitalia and, of course, complete sterility. The fact that marmoset monkeys do not develop the freemartin syndrome, even

Fig. 3.3. The consequences of placental vascular anastomoses. (a) When twin bovine fetuses of opposite sex share a common circulation as a result of fusion of placental blood vessels (arrowed), the ovaries of the female fetus become transformed into sterile testes, and her reproductive tract is partially masculinized. (From F. R. Lillie. *J. Exp. Zool.* **23**, 371–452 (1917).) (b) When twin marmoset monkey fetuses of opposite sex share a common circulation as a result of fusion of placental blood vessels (arrowed), which is almost invariably the case, the development of the female fetus is completely unimpaired.

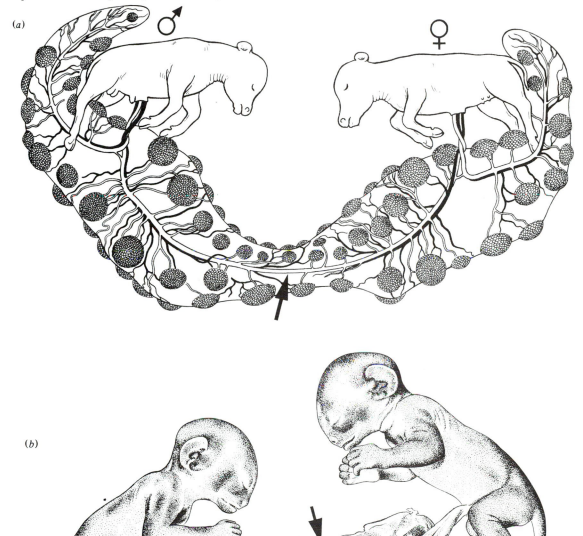

though their twin fetuses also share a common placental circulation, could be due to a lack of free circulating H-Y antigen in this species.

There is one other complicating factor about the H-Y antigen. Although it can apparently dictate the development of the somatic tissues of the indifferent gonad, resulting in the formation of Leydig cells, seminiferous tubules and Sertoli cells, it cannot bring about a functional sex reversal of the germ cells. We will discuss the intricacies of germ cell sex in a moment, but it is an inescapable conclusion that XX germ cells, or even XXY ones, cannot survive in a testicular environment, so that all XX males, even though H-Y antigen-positive, are sterile. The fact that germ-cell sex seems to be under a different genetic control mechanism that is entirely independent of the H-Y antigen would be in agreement with the observation that in males the immature germ cells are probably the only ones in the body that are H-Y antigen-negative.

Before we leave the male-determining role of the Y chromosome, there is one other Y-linked gene – maybe the only other one – that we must discuss. In mice, there is good evidence that a Y-linked gene determines the size of the testis, or more specifically the volume of seminiferous tubular tissue. It seems reasonable to suppose that this gene would be Y-linked in other species too; since the spermatogenic capacity of the testis needs to vary with the copulatory frequencies of different species, which are in turn a function of their different mating systems, a Y-linked gene would allow one to alter testicular size rapidly in response to selection without at the same time altering ovarian size. This topic is discussed more fully in Chapter 1, Book 8 in the First Edition, with particular reference to testis size in man and the great apes.

Having spent so long discussing the minute Y chromosome, what can we say about the much larger X chromosome, which is known to contain genes for at least 90 known characteristics in man? Susu Ohno has described in Book 6 of the First Edition how the X chromosome seems to have evolved by 'capturing' much of the genetic information formerly contained on the Y chromosome. In one species of rodent, the mole vole *Ellobius lutescens*, the process seems to have been so complete that the animal has been left with no Y chromosome at all. Both males and females have an XO karyotype, although presumably there must be some difference between the male X and the female X. Two other interesting variants are seen in species such as the mongoose, where the Y chromosome has become translocated onto one of the autosomes, and in some marsupials and bats, where it is the X chromosomes that have been translocated onto autosomes (see Book 6, Chapter 1).

The genes on the X chromosome of all mammals have been highly conserved during the course of evolution, so that if a character is sex-linked in one species, it tends to be sex-linked in all. But the evolution of such a large X chromosome, containing so much genetic information, has posed its own problems. How can the female survive with what amounts to a quadruple dose of X-linked genes on her two X chromosomes when the male, with his one X chromosome, has only a double dose? The answer is X-inactivation, the process by which the female suppresses almost all the genes on one of the two X chromosomes in somatic cells as a dosage-compensation device. The decision is normally a random one, made at a very early stage of embryonic development, and once a cell has decided whether it is the paternally or maternally derived X chromosome that is to be inactivated, all the progeny of that cell will behave in like manner. The inactivated X chromosome is heterochromatic, and [³H]thymidine labelling shows that it replicates its DNA later in the mitotic cycle than the other chromosomes.

There are a number of interesting exceptions to this rule of random X-inactivation. The first relates to the germ cells themselves. Although random X-inactivation seems to occur in the primordial germ cells, it is followed by reactivation, so that both X chromosomes are again functional in oocytes. As a corollary of this, it is interesting that the single active X chromosome of the male becomes inactivated early in meiosis. One animal, the creeping vole *Microtus oregoni*, even eliminates the X chromosome completely from its male germ cell line, a topic to which we will return when we discuss Germ Cell Sex.

The other exceptions to the rule of random female X-inactivation are also of great interest. In kangaroos, it seems that the paternal X chromosome is preferentially inactivated in most somatic tissues, and the same phenomenon seems to occur in the mule, where the X chromosome from the donkey father is more commonly inactivated than that from the horse mother. And in mice and rats, there is also evidence for paternal X-inactivation, but it is confined to the placenta. In humans, if one of the X chromosomes, should happen to be defective, it seems to be preferentially inactivated. And in females with more than two X chromosomes, such as XXX and XXXX women, or males with more than one X, such as XXY and XXXY men, all the surplus X chromosomes are inactivated. This dosage compensation device, coupled with an X chromosome that contains double the genetic complement, undoubtedly explains why it is that X chromosome monosomies and trisomies are compatible with life, whereas all autosomal monosomies, and most trisomies, are lethal (see Chapter 5). The absence of dosage compensation in the germ cell line also explains the infertility of XO females and XXY males in all species so far investigated.

Evidence of X chromosome inactivation can be seen with the naked eye. This is best illustrated in the case of the tortoiseshell cat, whose coat is a mosaic patchwork of black and yellow hairs. Black hair is produced by a dominant gene, *B*, and yellow by its recessive allele, *b*. The gene is sex-linked on the X chromosome, so if a female cat has inherited a black-containing X chromosome from one parent and a yellow-containing one from the other, random inactivation of the X chromosomes in her skin will give the typical mottled black and yellow appearance of the tortoiseshell.

Male tortoiseshell cats are understandably rare, and are of particular
interest, since they must have two genetically different X chromosomes.
They have been studied in some detail, and although a few are XXY in
karyotype and sterile, the commonest finding is that they are fertile, and
contain a mixture of XX and XY cells, or two populations of XY cells,
suggesting that the animals are true chimaeras formed, for example, by the
fusion of two fertilized eggs of different genotypes.

X-chromosome inactivation can also be seen at the cellular level, since
the heterochromatic X chromosome appears as a small, dense mass, the
sex chromatin or Barr body, in the nuclei of some cells, or as a 'drumstick'
extension to the nucleus in some polymorphonuclear leucocytes. In man,
it is also possible to make the Y chromosome visible in cells, because it
will fluoresce in ultraviolet radiation after staining with quinacrine. This
property enables us to distinguish X- from Y-bearing spermatozoa, and
thereby provides a ready check on techniques for the separation of X from
Y spermatozoa. Unfortunately, this brilliant Y fluorescence cannot be used
to distinguish X- from Y-bearing spermatozoa in any other species (see
Fig. 3.4). The ability to sex cells on the basis of sex chromatin has provided
us with a 'poor man's karyotype', and sex chromatin still has its uses,
for example in determining the sex of human fetuses from the appearance
of desquamated cells in amniotic fluid obtained at amniocentesis; it is

Fig. 3.4. Intranuclear sex
chromatin or Barr body
(arrowed) in a female nerve
cell (*top left*), and in two
female fibroblasts (*top
right*). One 'drumstick' is
seen in a female poly-
morphonuclear leucocyte
(*bottom left*). The
fluorescent Y chromosome
can be seen in the heads of
human spermatozoa
following fixation and
staining with quinacrine
(*bottom right*).

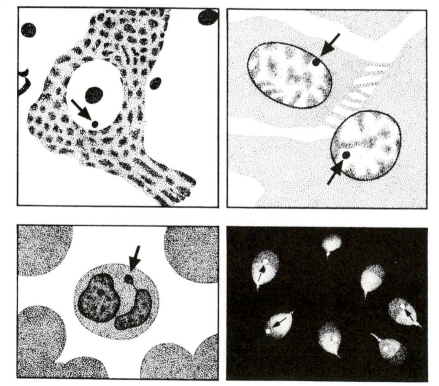

sometimes difficult to obtain mitotic preparations from such epithelial cells for a more definitive assignment of sex by karyotyping.

Before leaving the subject of the role of the X chromosome in the determination of sex, there is one more oddity that we must discuss – the XY female. Perhaps a good place to begin is with an obscure publication by two Finnish biologists, Kalela and Oksala. In 1966, they reported in *Annales Universitatis Turkuensis*, a journal that few people can ever have read, the amazing fact that in the wood lemming *Myopus schisticolor* certain females seem to be able to give birth only to female offspring. This paper caught the attention of a Swedish geneticist, Karl Fredga, then working in Lund, and he thought that it might just be worth devoting his summer holidays to catching some of these animals, in order to karyotype them. Unlike the common lemming, with its cycles of superabundance, the wood lemming is a shy, retiring creature that is not particularly abundant, or easy to trap. But Karl, a great naturalist in the Linnaean tradition, was successful, and how worthwhile that forsaken holiday proved to be. For the wood lemming, it turns out, has several classes of females in the

Fig. 3.5. Sex-determining mechanism in the wood lemming *Myopus schisticolor*. Certain females in the population possess a gene on their X chromosome (designated Ẋ) that can suppress the male-determining role of the Y chromosome. Thus there are three classes of females in the population; normal XX females, heterozygous ẊX females, and ẊY females. The sex ratios in the offspring of the latter two classes are skewed, to give 3♀:1♂ or 4♀:0♂, respectively. In the ẊY females, the Y chromosome is lost from most of the germ cells, and the single Ẋ doubles up by non-disjunction to give ẊẊ germ cells. All males have a normal XY karyotype, since the presence of an Ẋ chromosome is incompatible with the development of a male phenotype. (Adapted from K. Fredga, A. Gropp, H. Winking and F. Frank. *Hereditas*, **85**, 101–4 (1977).)

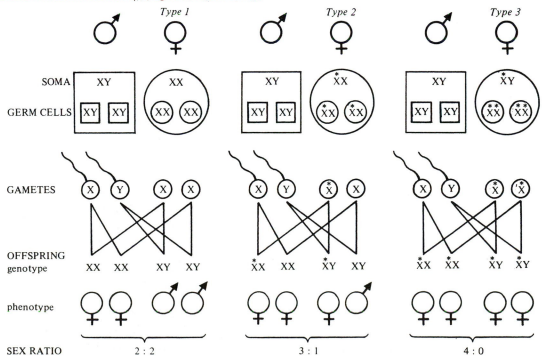

population. Some have a normal XX karyotype, and produce both male and female offspring, whereas others have an XY karyotype, and produce almost exclusively females. Further studies of these XY females showed that the Y chromosome was usually lost from the germ cell line, and that the single X chromosome doubled up by a process of non-disjunction to give normal XX oocytes in an individual whose somatic tissues were all XY. But how did the gonad escape the male-determining role of the Y in the first place? The most plausible explanation is that there must be an X-linked gene that suppresses the activity of the Y chromosome, and this idea was subsequently supported by the observation that there are indeed two morphologically distinct X chromosomes visible in the wood lemming's karyotype, the "suppressor X" (designated $\overset{*}{X}$) having smaller short arms and a distinct banding pattern on Giemsa staining. Putting the whole story together, it seems that all wood lemming males have a normal XY karyotype, but that there are three classes of females, normal XX, heterozygous $\overset{*}{X}X$, and $\overset{*}{X}Y$. The XX females, when mated to XY males, have the expected $2\male:2\female$ sex ratio in their offspring, whereas the heterozygous females give a $1\male:3\female$ sex ratio, and the $\overset{*}{X}Y$ females give almost entirely female offspring (see Fig. 3.5). Very occasionally, an $\overset{*}{X}Y$ female will give birth to a normal XY male, suggesting that a few $\overset{*}{X}Y$ oogonia survive, to produce the occasional Y-bearing oocyte.

More recently, Russian cytogeneticists have shown that the Arctic varying lemming *Dicrostonyx torquatus*, so-called because it changes its coat colour from grey in summer to white in winter, also has $\overset{*}{X}Y$ females in the population. However, they differ from the wood lemming because they produce numerous XY male offspring. The conclusion is that the varying lemming must have retained its $\overset{*}{X}Y$ oogonia, and dispensed with non-disjunction, to produce $\overset{*}{X}$- and Y-bearing oocytes. If a Y-bearing oocyte is fertilized by a Y-bearing spermatozoon, the embryo will perish, since at least one X chromosome is essential for survival in all mammals.

The reason for discussing wood lemmings and varying lemmings at such length is that these intriguing animals may help us to explain a hitherto baffling condition, the XY human female. If the X chromosome has been highly conserved during the course of evolution, as we believe, it seems likely that X-linked Y-suppressor genes will be found in animals other than the lemmings, and the human XY female is a case in point; perhaps she, too, is in reality $\overset{*}{X}Y$. However, there is an important distinction, because XY women are always sterile, with 'streak' gonads and an absence of secondary sexual characteristics, suggesting that their XY oogonia have not developed either of the survival strategies seen in the wood lemming or the varying lemming.

We are destined to hear a great deal more about XY females in the future, because they provide one clue to the male-determining role of H-Y antigen. Initial studies in XY wood lemmings and women showed that they were H-Y antigen-negative, in keeping with the postulated Y-suppressor role of the X chromosome, but now H-Y antigen-*positive* XY females have been discovered. Perhaps they remain female because they lack receptors for the antigen? Recently Dr Wolf and his colleagues in Freiburg have suggested that part of the short arm of the human X chromosome regulates expression of the H-Y antigen. They think that the structural gene for H-Y antigen is in fact autosomal; it is normally repressed by a regulator gene on the X chromosome, which is in turn repressed by the Y chromosome. So maleness is brought about by inhibition of an inhibition!

Gonadal sex

The most critical event in the translation of genetic sex into phenotypic sex is the determination of the type of gonad that is formed. Indeed, the sex determining genes exert their vital role in only a few cells in the developing genital ridge of the early embryo, and they have no effect on the development of sexuality in any other body tissue. True, there are other genes, both sex-linked (X-linked) and autosomal, that can have a profound effect on sexual differentiation, but they work at a later stage as modifiers of gonadal hormone action.

Perhaps the most striking illustration of the all-powerful effects of the gonads on sexual development comes from the rare recorded cases of monozygotic human twins, one of whom was a boy and one a girl. The explanation for this incredible situation is that soon after fertilization, one cell in a developing male embryo lost a Y chromosome. The embryo then divided into two, and as a result the tissues of both twins were a mosaic of XY and XO cells; in one twin, XY cells predominated in the genital ridge, causing it to develop as a testis, whereas in the other twin XO cells gave rise to an ovary. Subsequent sexual development of the twins along different paths was determined by the loss of a Y chromosome from a single cell in the early embryo.

Anne Grete Byskov has already described in Book 1 the anatomical processes by which the indifferent genital ridge develops into an ovary or a testis. Precisely how the male-determining gene or genes exert their effect is not known, but under their influence seminiferous cords develop; they become populated by male primordial germ cells, and subsequently canalized to form the seminiferous tubules. In the absence of male-determining genes, the gonad develops into an ovary, with nests of female primordial germ cells in the ovarian cortex. Once gonadal sex has been determined by the genotype, subsequent development is virtually immutable. Steroid hormones, for example, which can readily bring about a complete, functional gonadal sex reversal in some fish and amphibians (see Book 6, Chapter 1), are completely without effect on the gonads of eutherian mammals, although there is one report which suggests that oestrogen may be able to induce a partial sex reversal of the testis in the pouch young of a marsupial, the Virginia opossum. Probably the only example of gonadal sex reversal in eutherian mammals is that seen in the freemartin condition, which was described in the preceding section.

Perhaps the immutability of mammalian gonadal sex is a consequence of viviparity; Nature could not afford to have gonadal sex subject to the whim of steroid hormones, when the mammalian fetus is bathed by high concentrations of maternal and placental oestrogens. However, steroids will not cause gonadal sex reversal even in the chick embryo, where one might have expected to find an effect.

Birds are nevertheless of particular interest, since they can undergo a functional gonadal sex reversal. As we have said, in birds the female is the heterogametic (ZW) sex; an ovary will only develop in the presence of a female-determining gene or genes associated with the W chromosome. In the absence of this feminizing influence, the indifferent gonad will develop into a testis. Normally it is only the left gonad that develops into an ovary, and the right gonad remains vestigial and undifferentiated. If the ovary is removed surgically at an early stage of development, or occasionally if it is destroyed by disease, the right gonadal rudiment will reactivate and develop into a testis, and phenotypically the bird will change from female to male. This testis is capable of producing normal, fertile spermatozoa which are Z- and W-bearing, unlike normal avian spermatozoa which are all Z-bearing. In one recent experiment in which normal ZW hens were inseminated with Z and W spermatozoa from such a sex-reversed female, normal male (ZZ) and female (ZW) chicks were produced, but no WW ones, presumably because at least one Z chromosome is essential for survival, just as one X chromosome is essential in mammals.

Although this type of gonadal sex reversal is unlikely to help the commercial poultry breeder, since the sex ratio at hatching will be 1ZZ:2ZW:1WW (dead), it would be quite another matter if we could produce a sex-reversed mammal with XX testes and thus only X-bearing spermatozoa. The livestock industry would welcome any means of skewing the sex ratio in favour of females, whereas human societies usually have a preference for more males. But mammalian gonadal sex reversal may remain just a pipe dream, because of the constraints imposed by Germ Cell Sex.

Germ cell sex

We have already considered the genetic differences in sex-determining mechanisms that seem to separate the somatic tissues of the gonad from its germinal elements. For example, oocytes differ from somatic female cells in possessing two functional X chromosomes, and hence lack a dosage-compensation mechanism; the single X chromosome of male germ cells is apparently inactivated at meiosis, and furthermore male germ cells, unlike somatic cells, do not seem to express H-Y antigen.

We can get dramatic proof of the significance of these genetic differences between sex and soma by looking at two unusual mammals which highlight the point. First, there is the creeping vole from the State of Oregon, USA – *Microtus oregoni*. Susu Ohno showed some years ago that male creeping voles have lost the X chromosome from all their germ cells, whilst retaining it in the somatic tissues. Thus they produce two classes of spermatozoa, one lacking any sex chromosome at all, and one containing a Y chromosome. In the female creeping vole, on the other hand, the situation is reversed; the second X chromosome is absent in all the somatic cells, but the germ cells double-up the single X chromosome by non-disjunction, just as they did in the wood lemming, to become XX (see Fig. 3.6). The second animal is one of the bandicoots from Western Australia, *Isoodon obesulus*. This is a rabbit-sized marsupial that has been studied by David Hayman and Bob Martin from Adelaide. The somatic cells of both sexes are XO in this species, whereas the male and female germ cells have retained the normal XY/XX sex chromosome constitution. It seems that male somatic cells lose their Y chromosome, and female somatic cells their second X chromosome, during the pouch life of the young (see Fig. 3.7).

It has also been possible to get some interesting experimental evidence in mice that supports the concept of differing sex-determining mechanisms in germinal and somatic tissues. If artificial chimaeras are produced by

Fig. 3.6. Germinal and somatic sex in the creeping vole *Microtus oregoni*. The X chromosome is lost from the male germ-cell line so that two classes of spermatozoa are produced, one lacking any sex chromosome, and one containing a Y chromosome. Females have only a single X chromosome in their somatic tissues, but this doubles up by non-disjunction in the oogonia, to give normal X-bearing oocytes. (From S. Ohno, J. Jainchill and C. Stenius. *Cytogenetics*, **2**, 232–9 (1963).)

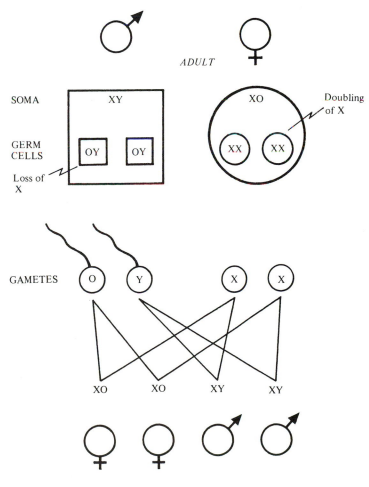

fusing XX and XY embryos, it becomes possible to study the fate of XX germ cells that migrate into a testis, and XY cells that migrate into an ovary; Richard Gardner discusses this subject in more detail in Chapter 6. The conclusion from such experiments is that XX germ cells are incapable of surviving in a testicular environment, and can never undergo complete spermatogenesis – they appear to perish in the attempt. However, one can very occasionally find an XY oocyte in the ovaries of female XX/XY chimaeras, which are themselves rather a rarity. The existence of XY oocytes is perhaps not so surprising when it is remembered that the wood lemming, and particularly the varying lemming, both have XY oocytes in their ovaries.

But creeping voles, bandicoots, and artificially produced chimaeric mice are perhaps rather exceptional examples if one is to try and prove the point that germ cell sex is a phenomenon that applies to all mammals. The most conclusive evidence about its all-pervasive nature therefore comes from

Fig. 3.7. Germinal and somatic sex in the bandicoot *Isoodon obesulus*. The male and female germ cells have a normal XY and XX karyotype, but the somatic cells of both sexes are XO. This is because the Y chromosome of the male, and the second X chromosome of the female, is lost from all somatic tissues during the pouch life of the young. (From D. L. Hayman and P. G. Martin. *Genetics*, **52**, 1201–06 (1965).)

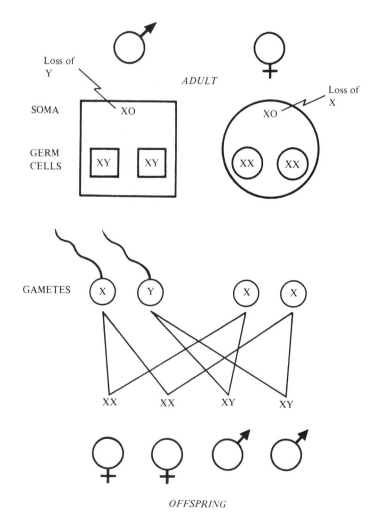

studying a series of spontaneously occurring genetic abnormalities in laboratory and domestic animals, and in man himself.

It will help us to understand these cases more clearly if we begin with some generalizations. The germinal and somatic tissues of the gonad have to interact with one another in order to produce functional gametes; thus a genetic abnormality that adversely affects the germ cells can ultimately have repercussions for somatic development, and similarly a genetic abnormality of the gonadal soma can influence germ cell development, and hence result in sterility. There is an interesting sex difference in this functional interrelationship, since the integrity of the germ cells in the ovary is essential for the secretion of hormones by the somatic cells, whereas in the testis the germ cells have little or no effect on hormone production.

Let us now consider a number of genetic disorders that illustrate these points, and we will begin by considering the male. We have already referred to the sex-reversal gene, *Sxr*, in the mouse when we were discussing genetic sex. This autosomal dominant gene causes the somatic tissues of the gonad of genetic females to develop into a testis, which secretes normal amounts of male sex hormones and thus brings about a complete transformation of the genetic female into a phenotypic male. A similar condition occurs in goats, under the influence of the autosomal gene *P* (*Polled*). However, in all these cases the XX germ cells are apparently unaffected by the *Sxr* or *P* gene, retain their integrity as oogonia, undergo meiosis to become oocytes, and ultimately perish in the testicular environment in which they find themselves (Fig. 3.8*a*).

A completely different story is seen if the *Sxr* gene is incorporated into the genetic background of an XO mouse. Normally, XO mice develop as females with reduced fertility. Not only do they produce many embryos that die *in utero*, but they also experience an accelerated rate of oocyte atresia, so that they become sterile at an early age. We will return to the reasons for this when we discuss XO humans, but for the moment let us concentrate on the consequences of sex-reversing the gonad of one of these XO females by incorporating the *Sxr* gene into its make-up. When this is done, the XO germ cells also seem to be sex reversed, and a few are able to undergo spermatogenesis (Fig. 3.8*b*) to produce motile spermatozoa, many of which are highly abnormal. Maybe this is because the germ cells lack a Y chromosome. But the significant point is that XO germ cells in the mouse can develop either into oogonia or spermatogonia, apparently depending on whether they find themselves in a testis or an ovary. This reinforces the view expressed at the beginning of this section that germ cell sex may be partly a consequence of the inability of the germ cell line to inactivate surplus X chromosomes.

Further support for this view is seen in individuals with an XXY karyotype. This condition has now been recognized in a wide range of animals, including mice, cats, dogs, sheep, pigs, bulls, and of course humans, where it affects 1 in 1000 newborn males. Such individuals subsequently develop the clinical stigmata referred to as Klinefelter's syndrome. But regardless of species, XXY individuals have small atrophic testes devoid of germ cells – usually. But there are occasional fascinating exceptions. Mary Lyon, the discoverer of the X-inactivation story, has shown that XXY mice have occasional seminiferous tubules in which spermatogenesis is taking place, and those particular germ cells have reverted to a normal male XY karyotype by loss of the surplus X. A similar situation seems to exist in the occasional Klinefelter's syndrome individual.

There is just one more condition we should discuss in the male that has

Fig. 3.8. (*a*) Oocytes developing in the testis of an 11-day-old sex-reversed female mouse (XX, *Sxr*/+). These will all have perished by about the sixth week of life. (From A. McLaren. *Nature, Lond.*, **283**, 688–9 (1980). (*b*) Spermatogenesis in the testis of a sex-reversed XO female mouse (XO, *Sxr*/+). The absence of a second X chromosome has made it possible for the germ cells to undergo sex reversal in the testicular environment in which they are located; XO germ cells in an ovary would normally give rise to oocytes. Courtesy of B. M. Cattanach.

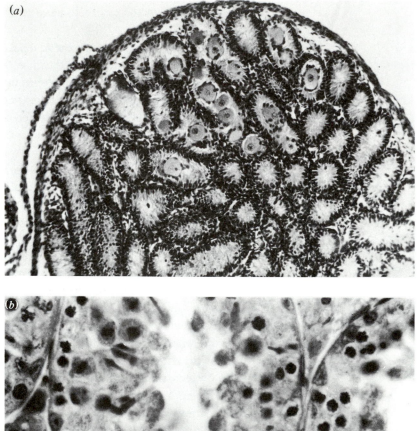

a bearing on germ cell sex, and that relates to the role of the Y chromosome. About 1 in 1000 newborn boys has an XYY karyotype, as a result of fertilization of an egg by a spermatozoon carrying two Y chromosomes; XYY mice are also known to occur. In both instances, testicular histology is abnormal, many tubules showing spermatogenic arrest. However, XYY individuals are fertile, and it seems that normal spermatogenesis can take place if the germ cells are able to lose the surplus Y chromosome. This is therefore another good example of a situation in which the somatic tissues can readily handle a surplus sex chromosome, in this case a Y, but the germ cells get into difficulties, apparently because of the presence of an unpaired Y at meiosis.

Turning to genetic disorders in the female, the most interesting is the XO condition, which occurs not only in mice, but also in horses, rhesus monkeys, and of course in humans. XO individuals exhibit a number of somatic abnormalities, referred to as Turner's syndrome. These include 'webbing' of the neck, coarctation (narrowing) of the aorta, renal abnormalities, an abnormal lymphatic system resulting in oedema, pigmented naevi (birth marks) in the skin, and a short stature. This may mean that the 'inactive' second X chromosome in female somatic cells is not as inactive as we had imagined. XO is in fact the commonest chromosomal abnormality in man, with an incidence at conception of about 1 per cent; but only one in 10 000 newborn females is XO, as over 99 per cent of XO embryos or fetuses will be aborted (see also Chapter 5). In humans, studies of the inheritance of the sex-linked blood group factor, Xg, have shown that in about 80 per cent of cases the single X is derived from the mother; thus the condition is usually caused by lack of a sex chromosome in the fertilizing spermatozoon.

We have already mentioned that XO mice have reduced fertility, an accelerated rate of oocyte atresia, and a premature onset of sterility. The same is true in humans, but because of the different time spans involved, the consequences are somewhat different. The ovaries of aborted XO human embryos look normal, and contain a normal population of germ cells. But the accelerated rate of oocyte atresia during fetal life means that by birth, there are few if any oocytes left. Without oocytes, there can be no Graafian follicles, and hence no oestrogen or progesterone secretion by the ovaries. The ovaries therefore degenerate into streaks of fibrous tissue, and in the absence of ovarian hormones there is little or no development of secondary sexual characteristics such as pubic hair or breasts, and menstruation never occurs. Very occasionally a few oocytes may survive until the time of puberty, and there are even on record one or two XO individuals who have actually had children; but these are exceptions that prove the rule. If oocytes lack two functional X chromosomes, they are adversely affected. And if germ cells disappear from the ovary, its somatic cells are incapable of secreting hormones, so the phenotypic characteristics of the whole of the rest of the body are ultimately affected. This must surely be the most dramatic example of the way in which germ cell sex can profoundly influence gonadal and hence phenotypic sex.

Hormonal sex

Once the gonad has become established as a result of the complex interplay of controlling factors that we have already discussed, the rest of sexual differentiation becomes much easier to understand. For a start, the hormones produced by the gonad depend on what the gonad looks like, and not on its genetic make-up. Thus a testis will secrete male sex hormones, regardless of whether it is genetically XY, XX, XXY, XYY, or XO. An ovary will secrete female sex hormones, but because it needs functional germ cells in order to do so, it does operate under severe genetic constraints.

It is also a general rule that the heterogametic sex is the dominant one, hormonally speaking. Thus in mammals, maleness is superimposed on the neutral female state by the action of male sex hormones. Castrate a male fetus, and it will develop as a phenotypic female; castrate a female fetus, and its development is unimpaired. In birds, where the female is the heterogametic sex, it is femaleness that is imposed on the neutral male state by the action of female sex hormones. Ovariectomize a hen, or a pheasant (and remove the undifferentiated right gonadal rudiment to prevent it developing into a testis), and the bird will develop the gaudy plumage of the male, although it will lack androgen-dependent male secondary sexual characteristics like a comb or spurs.

Fig. 3.9. The classical experiment of Alfred Jost, in which he grafted a piece of testis against one ovary of a rabbit fetus. The testis produced a local stimulation of the Wolffian duct and inhibition of the Müllerian duct, but the effect did not extend to the other side of the body.

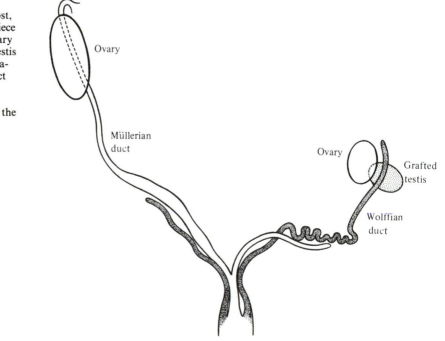

Ovary

Müllerian duct

Ovary

Grafted testis

Wolffian duct

In mammals, the testis produces different hormonal effects at different stages in the animal's life. In the fetus, its first task is to masculinize the adjacent genital ducts, and this is accomplished by local hormonal diffusion, rather than by generalized spread of hormones in the systemic circulation. Alfred Jost in Paris has been the pioneer in this field, and his classical experiments on rabbit fetuses, performed in the late 1940s, still provide the basis for our understanding of the subject. If a piece of testicular tissue from a male rabbit fetus is grafted against one ovary of a female fetus, the duct systems on that side of the body will be masculinized. The Wolffian (or mesonephric) duct, which ultimately forms the epididymis, vas deferens and seminal vesicles of the male, will hypertrophy, whereas the Müllerian (or paramesonephric) duct, which ultimately forms the fallopian tubes, cervix and part of the vagina of the female, will regress (see Fig. 3.9). If instead of a piece of testicular tissue a crystal of testosterone is used, the Wolffian duct will still hypertrophy, but the Müllerian duct will remain unaffected. This led Jost to suggest that the fetal testis must produce two different classes of hormones: androgens, which stimulate Wolffian development, and a postulated Müllerian inhibitory factor. Subsequent work by Nathalie Josso in Paris and Pat Donahoe in Boston has shown that this Müllerian inhibitory factor is a large molecular weight protein or polypeptide manufactured by the Sertoli cells in fetal life, and it should not be long before it has been completely isolated and characterized. Meanwhile, its terminology is already causing some confusion. Nathalie Josso likes to call it Anti-Müllerian Hormone; but it is not an anti-hormone, and strictly speaking we should not use the word 'hormone' until a substance has been completely identified.

So what are the androgens secreted by the fetal testis? There seems little doubt that the principal secretory product of the Leydig cells is in fact testosterone, and that this is responsible in its own right for stimulating the Wolffian duct to differentiate into the male internal genitalia. However, in certain target organs testosterone can be converted by the 5α-reductase enzyme system into another extremely potent androgen, 5α-dihydro-testosterone, and this seems to be responsible, in man at least, for masculinization of the external genitalia – formation of a penis and scrotum. Testosterone can also be converted by an aromatase enzyme system into oestradiol-17β, particularly in certain areas of the brain, and there is now good clinical and experimental evidence to show that oestrogen may be the biologically active form of androgen in the central nervous system in some species, a topic that we will discuss in the next section.

All these events can be summarized diagrammatically (Fig. 3.10). But before we leave the subject of hormonal sex, there is one final question we must ask; what makes the fetal testis secrete these hormones in the first place? Is it acting autonomously, or is it stimulated into action by either fetal pituitary or placental gonadotrophins? The evidence at the moment tends to favour autonomous secretion, since anencephalic or decapitated fetuses lacking a brain and pituitary undergo normal sexual differentiation of the internal genitalia, and this is true even of species that do not seem to produce a placental gonadotrophin. However, the fetal anterior pituitary has certainly assumed control of the gonads by late gestation.

Brain sex

Carl Pfeiffer and Geoffrey Harris were the first people to alert us to the fact that the brain itself must be a sexually differentiated organ. They grafted the ovaries of rats into the anterior chamber of the eye of males or females castrated in adulthood, and were able to show that it was only

Fig. 3.10. Hormonal activities of the mammalian fetal testis.

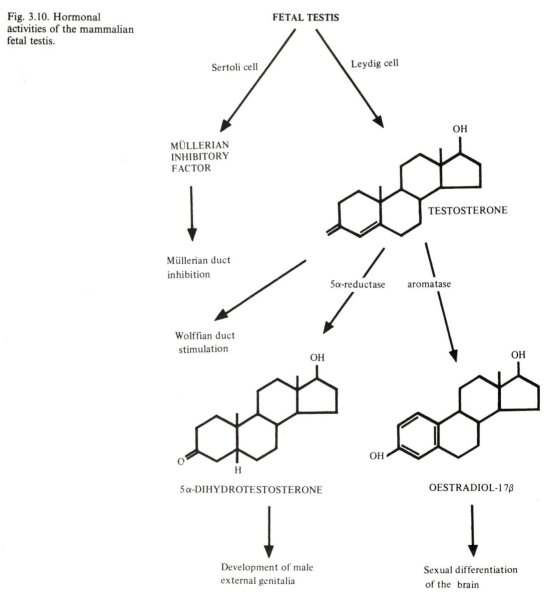

in the female host that the pattern of pituitary gonadotrophin secretion would allow normal ovulatory cycles; this suggested the existence of an important sex difference in the brain. Other workers showed that if newborn female rats were given an injection of testosterone in the first few days of life, they failed to show oestrous cycles after puberty, suggesting that their brains had in some way been masculinized. People then began to look for anatomical evidence of sex differences between the brains of male and female rats, and were able to detect subtle but significant differences in the neurones of the preoptic region. But it was Roger Gorski and his colleagues in Los Angeles who in 1978 provided the most dramatic evidence of this sex difference. They showed that the medial preoptic nucleus of the male rat is so much larger than that of the female that the difference can even be seen by the naked eye (see Fig. 3.11). Careful quantitative studies showed that the volume of this nucleus was eight times greater in males than in females, a difference that is apparently due to the presence of more large neurons. If male rats are castrated on the day of birth, the volume of the medial preoptic nucleus is greatly reduced; if newborn females are injected with testosterone, then their nuclear volumes are significantly increased. However, castration or hormone administration later in life has no effect.

It would be tempting to conclude that the anatomical difference in nuclear size must account for the different patterns of gonadotrophin secretion shown by male and female rats, but this does not seem to be the

Fig. 3.11. Sexual dimorphism in the size of the medial preoptic nucleus of the brain in the rat. (*a*) represents a saggital section, and (*b*) a transverse section. The male nuclear size is shown in heavy stipple, and the female nuclear size as an unshaded area within it. (Adapted from R. A. Gorski, J. H. Gordon, J. E. Shryne and A. M. Southern. *Brain Res.* **148**, 333–46 (1978).)

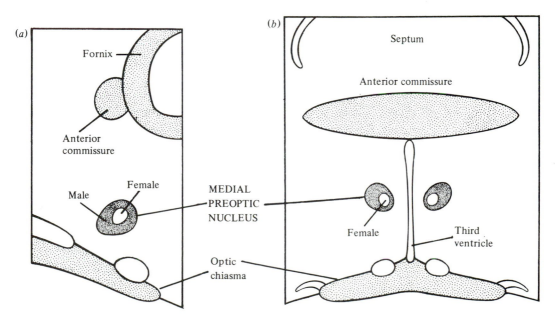

case – gonadotrophin-releasing hormone secretion is regulated by the arcuate and ventromedial nuclei, not by the preoptic nucleus (see Chapter 1, Book 7 in the First Edition). It seems more likely that the medial preoptic nucleus is concerned with the regulation of sexual behaviour, which is certainly sexually dimorphic. This is a topic we will return to in a later section.

How do sex hormones produce these anatomical changes in the brain? It seems that they stimulate the proliferation of nervous tissue directly, because if explants of the preoptic region of mouse hypothalamus are grown in culture, the addition of testosterone or oestradiol-17β will produce a spectacular outgrowth of neurites, whereas these steroids are without effect on explants of other areas of the brain, for example the cerebral cortex. Thus testicular steroids produced in fetal or neonatal life may be able to alter the 'wiring diagram' of certain areas of the brain, an organizational effect that can have profound consequences in later life. It will be exciting to see whether sexually dimorphic areas exist in the human brain, and if so, what functions they regulate.

We already have some biochemical evidence to suggest that there are differences in brain sex in other species. Injection of oestrogen can provoke a reflex discharge of LH from the pituitary in intact or castrated ewes, but not from intact or castrated rams. Exposure of the female fetus to androgens early in gestation will abolish this response, thus making ovulation impossible in later life. However, this biochemical test of 'brain sex' does not enable us to predict the type of sexual behaviour the sheep will show in adult life, which is perhaps not so surprising since gonadotrophin secretion and sexual behaviour are controlled by different areas of the brain.

It was natural to suppose that brain sex would be determined by testosterone itself, but doubts began to arise when it was discovered that you could masculinize the brain of a newborn female rat more effectively with oestradiol than you could with testosterone itself. The plot thickened when it was shown that 5α-dihydrotestosterone, a non-aromatizable androgen that cannot be converted into oestrogen, will not masculinize the brain, and Fred Naftolin put forward the hypothesis that testosterone may first have to be aromatized in the central nervous system to oestradiol-17β before it can exert its effects. In support of this idea, he demonstrated aromatizing activity in brain tissue in a variety of species, including man. Others showed that the rat brain contains oestrogen receptors, and that anti-oestrogens will effectively prevent the masculinizing effects of testosterone. So we must accept the fact that in the rat, brain aromatization does seem to be an important pre-requisite for androgen action. It remains to be determined whether this is true for all species, and whether it applies to all the effects of androgen in the central nervous system.

Phenotypic sex

We are now in a position to understand how the hormones we know about determine the individual's sexual phenotype. Before entering into a discussion of abnormalities, we must first understand the normal processes of sexual differentiation. Fig. 3.12 summarizes the development of the internal genitalia, and Fig. 3.13 the development of the external genitalia in the human. It can be seen how the genital tubercle develops into the glans penis of the male or the clitoris of the female. The genital or urethral folds fuse together in the male from the anus right up to the tip of the penis, thereby creating the penile urethra; you can still see the fusion line or raphé in the adult. The urethral folds in the male also form the scrotum. In the female, the urethral folds do not fuse, and come to form the labia majora on either side of the vaginal opening.

Fig. 3.12. Development of the male and female internal genitalia. In the male, the prostate and the bulbourethral glands are outgrowths of the urogenital sinus, and are not derived from the Wolffian duct. In the female, the lower part of the vagina is formed from the urogenital sinus. (Adapted from S. S. C. Yen and R. B. Jaffe. *Reproductive Endocrinology.* Saunders; Philadelphia (1978).)

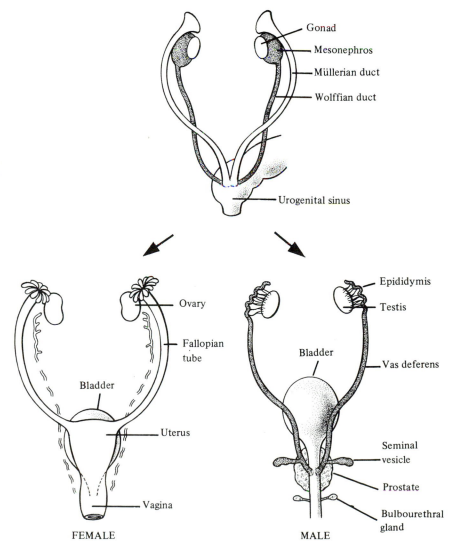

Gonad
Mesonephros
Müllerian duct
Wolffian duct
Urogenital sinus

Ovary
Fallopian tube
Bladder
Uterus
Vagina

FEMALE

Epididymis
Testis
Bladder
Vas deferens
Seminal vesicle
Prostate
Bulbourethral gland

MALE

One particularly interesting aspect of sexual differentiation that is often glossed over in textbooks is testicular descent (see Fig. 3.14). This can be divided into three stages: nephric displacement, transabdominal passage, and inguinal passage. Initially the testis is anchored to the caudal end of the abdominal cavity by the Wolffian duct and the caudal genital ligament, which connects up with a fibrous cord, the gubernaculum testis; the definitive metanephric kidney moves away from the testis by migrating anteriorly. Transabdominal passage of the testis is usually thought to be due to the pull of the gubernaculum, which is anchored in the region of the genital fold. Since the gubernaculum contains no muscle fibres, any traction it exerts must be by a process of differential growth. The final stage of testicular descent, through the inguinal canal and into the scrotum, is perhaps the least understood. Some have likened it to a herniation process, under the influence of increased intra-abdominal pressure, and others have suggested that the distal end of the gubernaculum swells up by inbibing

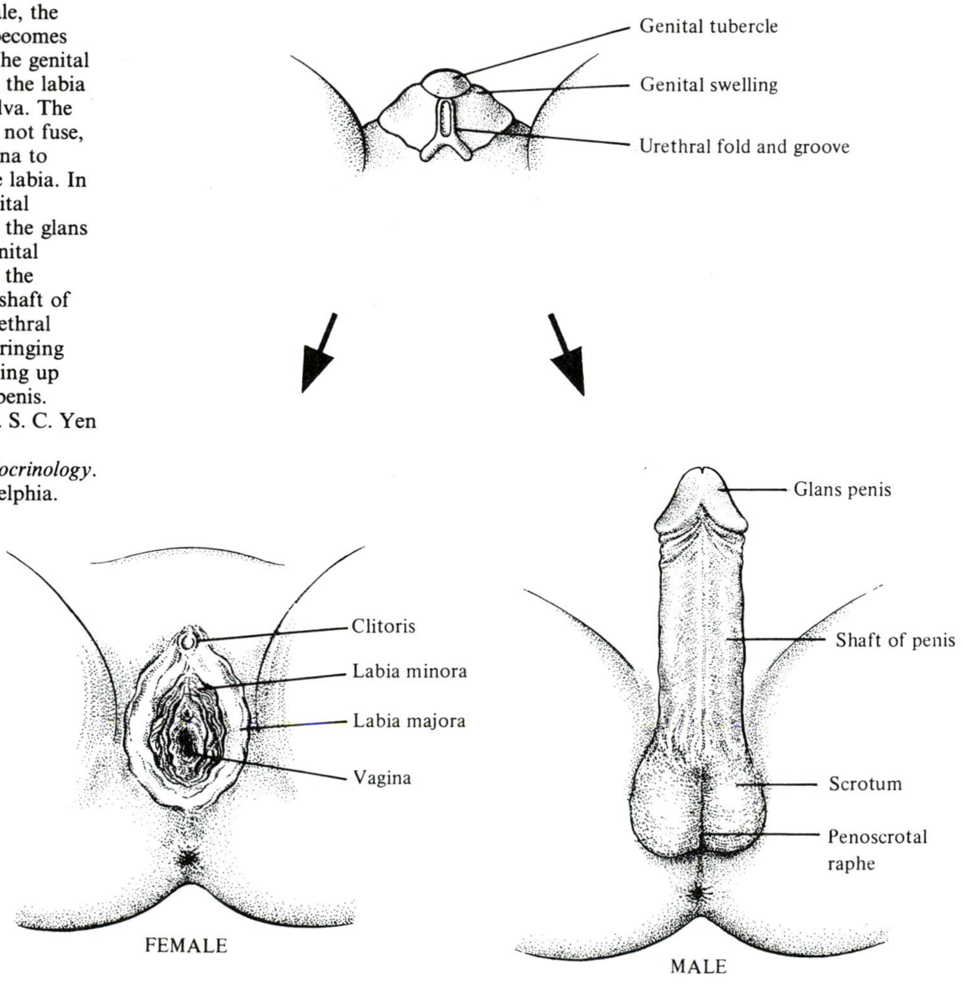

Fig. 3.13. Development of the external genitalia in man. In the female, the genital tubercle becomes the clitoris, and the genital swelling becomes the labia majora of the vulva. The urethral folds do not fuse, allowing the vagina to open between the labia. In the male, the genital tubercle becomes the glans penis, and the genital swelling becomes the scrotum and the shaft of the penis. The urethral folds fuse, thus bringing the urethral opening up to the tip of the penis. (Adapted from S. S. C. Yen and R. B. Jaffe. *Reproductive Endocrinology*. Saunders; Philadelphia. (1978).)

Genital tubercle

Genital swelling

Urethral fold and groove

Glans penis

Shaft of penis

Scrotum

Penoscrotal raphe

Clitoris

Labia minora

Labia majora

Vagina

FEMALE

MALE

fluid, and in so doing exerts traction on the testis, thus pulling it through the inguinal canal into the scrotum.

There can be little doubt that testicular descent is under hormonal control, and the generally accepted view has been that it is androgen-dependent. Although androgens are probably involved, there is reason to believe that Müllerian inhibitory factor may also play a part. In patients with the testicular feminization syndrome, an X-linked genetic defect makes all the target tissues totally androgen-insensitive, so that the external genitalia are female; the testes have nevertheless descended into the inguinal canal or the labia majora in 80 per cent of cases. And at the other extreme, in men with the persistent Müllerian duct syndrome, who either lack or are resistant to Müllerian inhibitory factor and hence have normal male external genitalia but retention of the uterus, the testes remain intra-abdominal.

In man, the final stage of testicular descent into the scrotum begins at about the eighth month of gestation and is completed at about the time of birth. Thus, in premature infants weighing less than 1500 g, 60–70 per cent will have undescended testes, but the figure has dropped to about 3 per cent in normal full term infants, and to 1 per cent by 1 year of age.

Fig. 3.14. Human testicular descent. This begins at about the sixth week of gestation, but the process is not completed until about the time of birth. (Adapted from J. Langman. *Medical Embryology*, 3rd edn. Williams and Wilkins; Baltimore. (1975).)

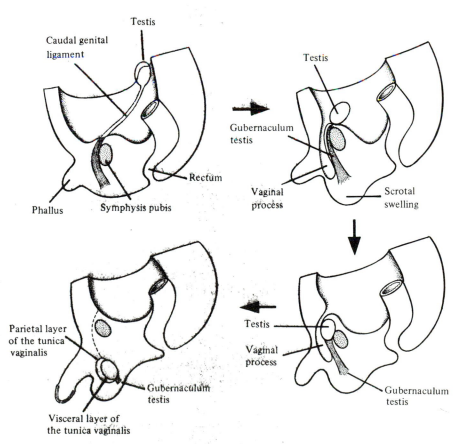

The incidence of undescended testes in adult men is about 0.4 per cent. The condition is often unilateral, and the cryptorchid testis has usually completed its transabdominal passage and is to be found lying in the region of the inguinal canal. The longer the testis remains undescended after the first or second year of life, the greater the chance of developing a testicular tumour, and the less chance that it will ever be capable of producing spermatozoa. But it remains a 'chicken-and-egg' situation, since we do not know whether the testis fails to descend because it is abnormal, or whether it becomes abnormal because it fails to descend.

The sex differences in adult phenotype do not become fully established until the time of puberty, and in no species has puberty been studied in

Fig. 3.15. Sequence of pubertal changes in boys and girls. (Adapted from W. A. Marshall. *J. Biosoc. Sci.*, *suppl.* **2**, 31–41 (1970).)

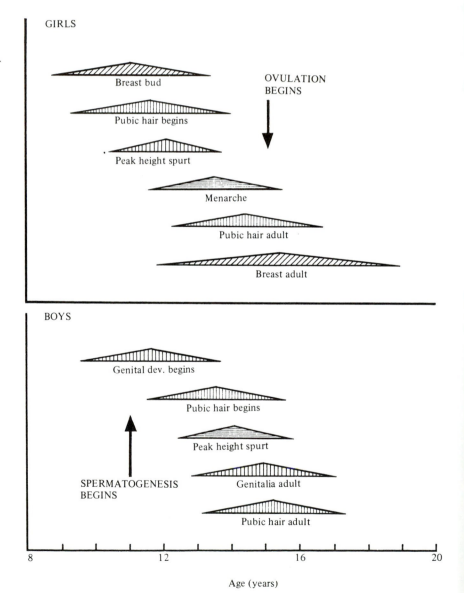

such minute detail as in man (see Chapter 1, Book 8 of the First Edition). Fig. 3.15 summarizes the sequence of events. In girls, increasing pituitary gonadotrophin secretion causes waves of follicular development in the ovaries towards the end of the first decade of life, and the first target organ to respond to the oestrogen so produced is the breast. But the ovaries also produce androgens, and rising ovarian androgen secretion, eked out maybe by some adrenal androgens, stimulates pubic hair development. The combined effect of these very low levels of oestrogen and androgen is initially to stimulate division of the epiphyseal cartilages in the long bones, so that there is a sudden spurt in growth. As oestrogen secretion rises still further when larger and larger follicles begin to form, the endometrium becomes stimulated, and eventually menstruation occurs in response to oestrogen withdrawal at the end of one of the cycles of follicular development. This first menstrual period (menarche) usually precedes the first ovulation by a year or so, and by the time ovulation eventually does occur, the phenotypic development of the girl is complete. She has attained her adult height, since elevated oestrogen levels result in epiphyseal closure and hence a cessation of growth; her breasts are fully developed, and she has an adult pattern of pubic hair development.

Pubertal development in boys follows a somewhat different course (Fig. 3.15). The first change in appearance is testicular enlargement, and this is due to increased pituitary gonadotrophin secretion stimulating the enlargement of the seminiferous tubules and the onset of spermatogenesis. Thus boys are potentially fertile at the very beginning of their pubertal development, whereas girls only become fertile at the very end of theirs. Gradually, the boy's enlarging testes begin to secrete sufficient androgen to cause penile enlargement and pubic hair development, and then to stimulate epiphyseal division and produce a growth spurt. These changes are completed in the mid to late teens, by which time the testes will have reached full size, and high androgen levels will have brought about epiphyseal closure, and the adult pattern of penile and pubic hair development.

If this is the normal course of development, we can now begin to understand the abnormalities that occur in a variety of clinical conditions. Let's begin by considering the abnormal appearance of individuals castrated in childhood, in whom secondary sexual characteristics do not develop. Fortunately, such individuals are extremely rare! However, we do have some information on boys castrated in childhood as the result of the activity of an obscure middle European sect, whose (intact) elders believed that childhood castration was a passport to Heaven. Iwan Gregor was such an individual, castrated at 5 years of age, and photographed at the age of 24 (Fig. 3.16*a*). There is some pubic hair, presumably as a result of adrenal androgen secretion, but there is no penile development. The excessively long arms and legs – the so-called eunuchoid characteristics – are due to failure of epiphyseal fusion, and hence continued slow long bone growth. Many eunuchs show signs of breast development (gynaeco-mastia) in later life, presumably because there is no androgen to counteract the low levels of adrenal oestrogen.

Patients with Klinefelter's syndrome (XXY) are not unlike eunuchs in a number of ways (Fig. 3.16b). As already mentioned, they have small testes and no surviving germ cells. For reasons that we do not fully understand, the testes usually secrete low levels of testosterone, and this may be reflected in the small size of the penis, and a sparse distribution of pubic, axillary, body and facial hair. There is often increased growth of the long bones, so they tend to be rather tall, and gynaecomastia is a common finding.

We have no information on girls ovariectomized in childhood – one hopes the operation has never been performed – but XY individuals with

Fig. 3.16. (*a*) A man, aged 24, who had been castrated at the age of 5. Note excessive growth of t'ie arms and legs due to failure of epiphyseal fusion, scanty pubic hair and minute penis. (*b*) A man with XXY Klinefelter's syndrome. Note small penis, sparse pubic hair, absence of body hair, and some breast development (gynaecomastia). The testes are small, and there is no spermatogenesis since XXY germ cells cannot survive in a testicular environment. (*c*) A woman with XY primary gonadal dysgenesis. It seems likely that an X-linked gene has suppressed the male-determining role of the Y chromosome, so the gonads develop as ovaries which degenerate into 'streaks', although the uterus and vagina are normal. Note scanty pubic hair in the absence of ovarian androgen, lack of breast development and tall stature in the absence of ovarian oestrogen. She is amenorrhoeic. (*d*) A woman with XY testicular feminization. There is no pubic hair, and the breasts are well developed in response to the unopposed testicular oestrogen secretion. The testes are in the inguinal region. They have produced normal amounts of Müllerian-inhibitory factor, hence there is no uterus present and therefore no menstruation; the vagina is short and blind-ending, being derived solely from the urogenital sinus. (*e*) A woman with XO Turner's syndrome. The excessively short stature and 'webbing' of the neck are probably due to the loss of the second X chromosome. XO oocytes undergo an accelerated rate of atresia and the ovaries degenerate into 'streaks', secreting no hormones. Note absence of pubic hair and failure of breast development. She is amenorrhoeic.

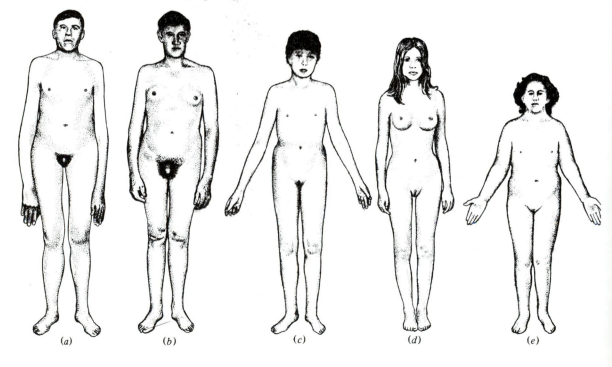

(*a*)   (*b*)   (*c*)   (*d*)   (*e*)

streak gonads (pure gonadal dysgenesis) can give us some clues as to what they might have looked like (Fig. 3.16c). You will remember that such individuals are thought to be the human equivalent of wood lemmings, with an X-linked Y-suppressor gene, so that although they have a male karyotype, the indifferent gonads are not transformed into testes, and they develop as phenotypic females. However, unlike the wood lemming, their germ cells do not survive, so their ovaries are non-functional. XY pure gonadal dysgenesis patients tend to be tall, with long limbs, because there is no oestrogen to produce epiphyseal fusion. As expected, they have no signs of breast development, very scanty pubic and axillary hair, and are amenorrhoeic. In contrast, XO gonadal dysgenesis patients (Turner's syndrome) are of short stature, which may be a reflection of their other cardiovascular and lymphatic abnormalities (Fig. 3.16e).

The above conditions are all a result of deficiencies in hormone production. There is another interesting group of conditions arising from abnormalities of hormone action, and of these by far the most spectacular is the testicular feminization syndrome, which we have referred to previously (Fig. 3.16d). This condition is due to an X-linked gene which causes an absence or an instability of the cytoplasmic androgen receptor protein in all cells throughout the body. In the absence of this receptor, neither testosterone nor its biologically active metabolite, 5α-dihydrotestosterone, can exert any effect; however, the cytoplasmic oestrogen receptor protein is unaffected. Phenotypically, human testicular feminization patients are very definitely female in appearance, often voluptuously so. They have an XY karyotype and testes which may be palpable in the inguinal region or in the labia majora. The testes are devoid of germ cells, presumably because of their cryptorchid nature, but have evidently produced Müllerian inhibitory factor in fetal life since these individuals lack Fallopian tubes, uterus, cervix, and the upper part of the vagina. Menstruation is thus impossible. Although the blood testosterone levels are in the male range, there are no peripheral signs of virilization, and a complete absence of pubic and axillary hair. There has been sufficient oestrogen production from the testes to bring about full mammary development, and it is interesting that such individuals also exhibit a 'male' pattern of pituitary gonadotrophin secretion, presumably because that centre in the brain was masculinized by oestrogen in the normal way. However, all other aspects of behaviour are female, and many of these women marry, have normal intercourse, adopt children, and lead a perfectly normal life as a woman.

Testicular feminization patients pose some ethical problems for the clinician. They are usually first picked up in gynaecology clinics, complaining of primary amenorrhoea. Because of the very real danger that their cryptorchid testes will develop into malignant tumours, it is necessary to remove them. The operation cannot be performed without the patient's consent, and castration is a shattering prospect for an attractive young girl contemplating marriage. But it is probably desirable to conceal the true diagnosis, because it would be an even greater psychological shock for the girl to learn the whole truth, namely that genetically, gonadally and hormonally she is male.

Another interesting abnormality of hormone action is the 5α-reductase deficiency, which was first discovered in certain families in the Dominican Republic. The condition appears to be due to an autosomal recessive gene, and it is only apparent in males. Affected individuals with a normal XY karyotype are born with female external genitalia, and are christened as girls, although internally they have a normal male reproductive tract, with testes that have migrated into the labia majora. The inability of peripheral tissues to convert testosterone into 5α-dihydrotestosterone has resulted in failure of development of the penis and scrotum. However, at puberty sufficient androgen is produced to bring about phallic enlargement, and many of the individuals spontaneously change name and sex to become boys, and may even get married to girls. This is a striking example of how society assigns gender on the basis of phenotypic appearance, but an individual's own sexuality is also determined hormonally, and conflicts can arise if phenotypic and hormonal sex are not congruent. We will return to this topic when we discuss Behavioural Sex.

A third type of hormone defect that may alter the human sexual phenotype, is where one gets abnormalities of hormone secretion. The commonest example of this is congenital adrenal hyperplasia (the adreno-genital syndrome). This condition is genetically determined, and results from the loss of one or other of the enzymes in the adrenal cortex that are responsible for synthesizing the principal adrenal hormones like aldosterone, corticosterone and cortisol. The activity of the adrenal cortex is normally controlled by the secretion of adrenocorticotrophic hormone (ACTH) from the anterior pituitary, and this in turn is regulated by a negative feedback of cortisol on the hypothalamus. If the adrenal cannot produce adequate amounts of cortisol because of the enzyme defect, then the pituitary will attempt to compensate by secreting an increased amount of ACTH, which in turn causes excessive enlargement of the fetal adrenals. Now the adrenal gland normally secretes only trace amounts of sex hormones, particularly androgens, but in congenital adrenal hyperplasia, this androgen secretion may reach significant proportions, and can cause extensive masculinization of the external genitalia of baby girls. The clitoris enlarges into a penis-like structure, and the labiae of the vulva may fuse to give a type of scrotum. Plastic surgery is required to restore these infants to their true phenotypic sex, and treatment with adrenal corticoids to make up for the cortisol deficiency. Baby boys suffering from congenital adrenal hyperplasia may show the 'Infant Hercules' syndrome, and reach puberty when only a few months or years old; but fortunately, they do not require plastic surgery.

There are also on record a number of girls whose genitalia were extensively masculinized during fetal life because their mothers were treated with synthetic progestational steroids in the mistaken belief that this might avert threatened or prevent habitual abortion. Unfortunately, some of the progestogens also possessed considerable androgenic activity, and crossed the placenta to masculinize the external genitalia of any female fetus.

The degree of masculinization produced by exogenous androgens depends not only on the dose, but also on the stage of gestation at which they are administered. We have shown in the sheep (gestation length = 150 days) that exposure of the female fetus to testosterone between days 40 and 50 will cause complete masculinization of the external genitalia, but treatment after day 50 will produce only a partial effect; development of the ovaries and uterus is of course unaffected. Thus days 40–50 appear to represent the 'critical period' for complete masculinization of the genital tubercle; it is interesting that the first histological evidence of gonadal differentiation in the sheep fetus is to be seen on day 35, and the external genitalia of the male normally begin to differentiate on day 45. Our information is not quite so precise in man, but we do know that gonadal differentiation occurs at about the 50th day of gestation, and the male external genitalia start to differentiate at about day 65, a process which is complete by about day 90.

Another aspect of phenotypic sex that we must mention is the true hermaphrodite. We still do not understand the cause of this rare condition; most true hermaphrodites are genetically **XX**, and because testicular tissue is present, the external genitalia usually show some signs of masculinization, although breast development may also occur at puberty. Problems arise internally because of the local nature of hormone action in fetal life. If testicular tissue is confined to one side of the body (usually the right), development of the Wolffian duct and inhibition of the Müllerian duct will be confined to that side of the body too. On the ovarian side, a Fallopian tube and uterus will persist, and the uterine lumen may connect with the urethra, to open into the penis. There are cases on record of true hermaphrodites starting to menstruate and ovulate at puberty, with the menstrual discharge passing out through the penis. And in pigs, where true hermaphroditism occurs not uncommonly, the animals may be fertile in the uterine horn adjacent to the (left) ovary, in spite of having a testosterone-secreting testis and a masculinized reproductive tract on the other side of the body.

At the beginning of this chapter, we made the sweeping statement that there is hardly a tissue or an organ of the body that does not show some form of sexual dimorphism, and yet so far all our discussions about phenotypic sex have been devoted to the spectacular changes in the anatomical appearance of the reproductive tract. So let us conclude with a brief account of some of the other tissues that are sexually dimorphic.

In mice, the kidneys and the submaxillary salivary glands are morphologically, histologically and biochemically different in males and females, and these differences are androgen-dependent. The renal changes may be concerned in part with the excretion of a pungent pheromone, smelling like acetanilide, in the urine of male mice; humans can even sex mice at a sniff. This pheromone is important for the control of territorial and aggressive behaviour in male mice. Sex differences in the submaxillary salivary gland are at first sight unexpected, as is the secretion of large amounts of nerve growth factor by the salivary gland of males. Perhaps this is an adaptation to wound healing; males may derive considerable benefit by retiring to lick their wounds after an affray.

It would be hard to imagine a less sexy organ than the liver, but in rats there are distinct sex differences in the liver's metabolism of certain steroid hormones and drugs, and these differences are androgen-dependent. Then of course there are the androgen-dependent differences in the volume and distribution of skeletal muscles, in skin thickness, and in tooth growth and hair distribution in many species, including our own. Even the larynx is sexually differentiated; Ebo Nieschlag and his colleagues in West Germany have shown that men who sing bass are taller, heavier, more athletic in build, and have a higher blood testosterone level than those who sing tenor. Finally, even life itself is sexually determined; men die sooner than women, but their life-spans can be increased by about 14 years following castration. So sexuality really does pervade the whole body in one way or another.

Sexual selection

We have now seen a whole array of secondary sexual characteristics by which the male can differ from the female. All these characteristics are androgen-dependent, but why did they evolve in the first place?

It was Charles Darwin who provided the answer in his book *The Descent of Man, and Selection in Relation to Sex*, published in 1871. He drew a distinction between Natural Selection, or the struggle of individuals of all ages and both sexes just to survive, and Sexual Selection, which he defined as 'the success of certain individuals over others of the same sex, in relation to the propagation of the species'. Sexual selection is discussed in greater detail in Book 6, Chapter 4 and Book 8, Chapter 1 in the First Edition. Briefly, it can be divided into two components: Intrasexual Selection, or the competition among members of the same sex, usually the males, for access to females, and Intersexual Selection, or the competition among members of one sex to attract members of the opposite sex. The extent to which sexual selection has operated to produce somatic dimorphisms is dependent on the mating system. In polygynous mating systems, where one male defends a group of females against the incursions of rival males, as in red deer or gorillas, competition among males is intense. Thus it is in polygynous species that we see the greatest degree of dimorphism of bodily characteristics related to inter-male combat, such as body size, muscular strength, and weapons of offence like antlers or canine teeth. At the other extreme, in monogamous mating systems where the male and female pair for life, inter-male combat is much reduced, so that there may be no visible male secondary sexual characteristics.

The lesson to be learned from all this is that it is the mating system that determines the extent to which secondary sexual characteristics develop at

Fig. 3.17. (*a*) The spotted hyaena *Crocuta crocuta*. (*b*) Ventral view of the external genitalia of an adult male spotted hyaena, showing the penis within its prepuce, and the scrotum and anus. (*c*) Ventral view of the external genitalia of a pubertal female spotted hyaena, showing the clitoris within its prepuce (indistinguishable in size, shape and location from the penis of the male), and the 'false' scrotum that is smaller than the male's since it does not contain any gonadal tissue. ((*b*) and (*c*) from L. Harrison Matthews. *Phil. Trans. Roy. Soc.* **B230**, 1–78 (1939).)

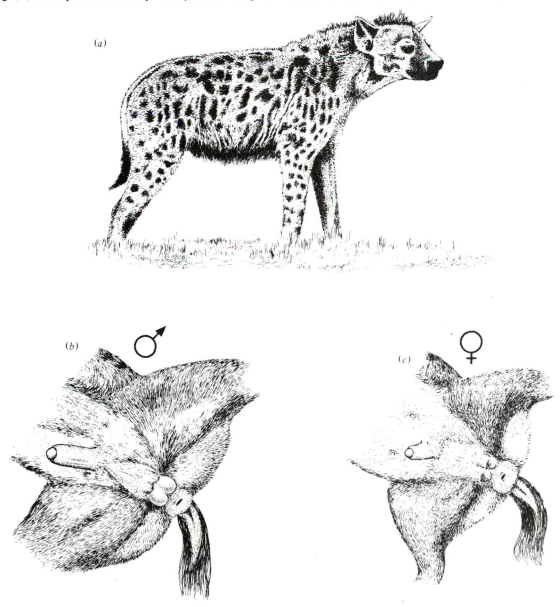

(*a*)

(*b*) ♂

(*c*) ♀

puberty. And what determines the mating system? A species has to develop a reproductive strategy that fits in with its lifestyle, and it may well be the female, not the male, who has the greatest say in the matter. She has the greatest energy investment in reproduction, and so she is the limiting resource. If she lives in an area where food is abundant, she can afford to share her home range with a male all the time, and so lifelong monogamy may be the result. If food is scarce, it is advantageous for her to live independently of the male except at mating time; hence the seasonal polygyny of the stag. We cannot begin to understand why sexual dimorphisms have developed until we have understood the animal's mating system in the wild.

The most dramatic illustration of this point is seen in the case of the spotted hyaena *Crocuta crocuta*. The ancients used to think that the hyaena was an hermaphrodite that could change its sex at will; this is understandable if we look at the external genitalia of the male and female, which are virtually indistinguishable (Fig. 3.17). The female has a false scrotum, which is empty, a prepuce, and within it a clitoris that is almost identical in size and shape to the penis of the male, and highly erectile. Since the vagina is totally enclosed by the clitoris, just as the male urethra is enclosed by the penis, the processes of parturition and copulation are indeed amazing.

But how has this masculinization of the female come about, and why? The 'how' is easy to answer; the ovaries of the female hyaena secrete testosterone, so that the peripheral blood levels are almost the same as those of the male. In order to understand why it has been necessary for the female to be so male-like, we need to look at the animal's behaviour. The spotted hyaena is a Jekyll and Hyde character, leading a communal existence and being a cringing scavenger by day, but a deadly killer by night; it seems probable that the female has had to protect herself and her young against the possibility of attack by the male. Thus she has become larger than the male, and has indulged in male mimicry as a defence strategy. When members of a clan meet after a temporary separation, there is an elaborate greeting ceremony in which genital inspection plays a vital role, and probably averts any aggression; the male erects his penis, and the female her clitoris during this display. The eerie chattering laughter, the loping gait, and the fearsome strength of those massive jaws make the spotted hyaena an animal to be feared and respected; but its incredible reproductive anatomy has much to teach us about the way in which sexual selection provides the ultimate explanation for sexual phenotype.

Behavioural sex

The whole subject of sexual behaviour is discussed at greater length in Book 4, Chapter 5 in the Second Edition, and Book 8 in the First Edition deals exclusively with human sexual behaviour. So here we will confine ourselves to a few of the salient points.

Clearly, sexual behaviour is just as dimorphic as any other phenotypic characteristic, and you might imagine that the pattern of behaviour was a direct response to stimulation by the different types of sex hormones produced by the testis or the ovary. It is certainly true that castrated female guinea pigs, hamsters, rats, mice, sheep, dogs, ferrets, red deer, or rhesus monkeys will not show oestrus, and that oestrogen injections, maybe in combination with progesterone, will restore oestrous behaviour within a day or two. The males of these species, and man himself, also show steadily declining levels of sexual desire in response to castration, although it may take months or years before it is lost completely; testosterone implants will restore it, but again there is usually a time lag of a week or more before the effects become obvious. The restoration of libido does not seem to be dose-dependent, since increasing the amount of hormone replaced will not increase the amount of sexual behaviour beyond the pre-castration level.

It cannot be the chemical nature of the sex hormone itself that is responsible for this sex-specific behaviour, since female sex hormones are notoriously unsuccessful at inducing female sexual behaviour in intact or castrated males of any species. Frank Beach, one of the pioneers in this field, has proposed that testicular androgens exert two different organizational effects in fetal or early neonatal life: 'defeminization', characterized by an inability to show female-type sexual behaviour in adulthood, and 'masculinization', or the enhancement of male-type behaviour. In species that give birth to immature young after a short gestation period, like rats, mice and hamsters, the critical period for masculinization and defeminization is around the time of birth, whereas in species like the guinea pig or sheep that give birth to mature young after a long gestation, the critical period occurs during fetal life. It is important to point out that in many species normal females will occasionally show male-type behavioural patterns, and, less commonly, normal males may sometimes show female-type behaviour, suggesting that the normal female shows traces of masculinization, and the normal male is not completely defeminized.

This masculinization–defeminization hypothesis could explain why it is that the castrated ram and stag have their male libido restored by oestrogen, and why the castrated ewe and hind have their oestrous behaviour restored by testosterone: this may be the only type of behaviour that they can exhibit in response to the activational effect of *any* sex hormone. It should also be remembered that if oestrogen is the biologically active form of androgen in the central nervous system, the only way of producing dimorphic sexual behaviour in response to a shared sex hormone is to produce dimorphisms in the neural substrate on which the hormone acts.

There are reasons for believing that androgens may also act in their own right to induce some types of behaviour, and that some behavioural responses do not depend on prior sexual differentiation of the brain. Take, for example, social aggression. We have studied this in free-living wild red deer, which have the advantage of having their sexual behaviour confined to the autumn months of the year, so that for the rest of the time it is possible to study social behaviour in isolation from any sexual context. Testosterone implants increase the social aggressive behaviour of the stag and hind (provided that she is not pregnant) in a dose-dependent manner, so that the implanted animal rises in dominance within the social hierarchy (Fig. 3.18*a*), and then fall again as the effect of the implant begins to wear off. This behaviour is *not* mimicked by oestrogen; an oestrogen implant that is releasing sufficient hormone to restore full sexual behaviour to a castrated stag has no effect on the social aggression in the rut or at other times of the year, and oestrogen does not appear to influence the social aggression of the hind.

Another important point to consider is the time-course over which the hormone acts. If a non-pregnant hind is implanted with testosterone, her initial behavioural response is to show oestrus within 1–2 days. But gradually this exclusively female type of behaviour gives way to one in which more and more male components are expressed, ultimately producing an animal that is still in oestrus in the sense that she is attractive to the male, but that is so aggressive that she will attack him if he attempts to mount her. As her male sexual behaviour develops still further, she will mount and attempt to copulate with all and sundry, including the human observer (Fig. 3.18*b*)!

A similar transition from female- to male-type sexual behaviour with time is seen in hinds or ewes implanted with oestrogen. This probably explains why the dairy farmer has so much trouble with his cows with cystic ovaries. Because they secrete low levels of oestrogen for long periods of

Fig. 3.18. Wild red deer hind, chronically implanted with testosterone in adult life, showing (*a*) aggression towards the observer, and (*b*) copulation with the observer!

(*a*)

time, they gradually begin to show more and more male-type 'bulling' behaviour, mounting all the other cows, and bellowing like a bull.

The fact that prolonged exposure to androgen or oestrogen can produce male sexual behaviour in the adult female suggests that whilst 'defeminization' is confined to fetal or early neonatal life, 'masculinization' can also occur during adulthood. The only reason that females do not normally show male sexual behaviour may be that, unlike the male, they are only exposed to elevated levels of sex hormones for brief periods of time whilst a follicle is developing in the ovary.

So far, we have discussed behavioural effects that are produced only in the presence of the hormone. However, there are also organizational effects of sex hormones in fetal or neonatal life that can permanently influence behaviour in adulthood even in the absence of hormones. The pattern of urination behaviour is a good case in point. Dogs cock their legs against vertical objects to scent mark, whereas bitches squat to urinate. Rams urinate without any noticeable postural change, whereas ewes arch their backs and crouch with their hind legs. These characteristic male patterns of urination in the adult are induced by testosterone acting in late fetal life in the case of the ram, or in late fetal and early neonatal life in the dog, and the behaviour will still develop and persist even if the animal is castrated after the first month of life.

These differences in urination posture may serve as long-range sexual signalling devices; since oestrous pheromones are excreted in the urine, it may be important for the male to be able to recognize a urinating female

(b)

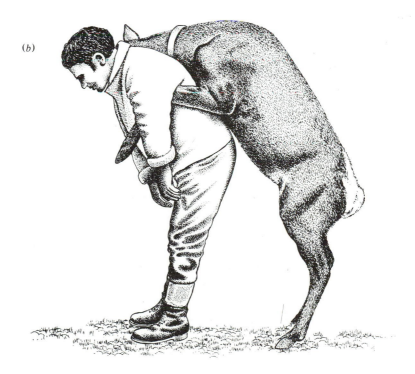

at long range so that he can go and inspect the spot, showing the male-specific *flehmen* response as the odour of the urine is exposed to his vomeronasal organ for sexual analysis. Man has also evolved a highly sexually dimorphic urination posture, which could be in part hormonally determined, and have social significance as a long-range visual signal.

Another possible example of an early organizational effect of androgens on behaviour is seen in primates. If female rhesus monkey or human fetuses are exposed to androgens, they will show male-type patterns of play behaviour, described as 'rough-and-tumble' play, during childhood. Although we cannot be certain that this behaviour will develop in the complete absence of hormones, it is certain that very little sex hormone is present prior to puberty when the behaviour is most evident.

We are unsure of the validity of the masculinization–defeminization hypothesis in primates, since androgenization of the female never seems to produce defeminization. When discussing Brain Sex, we saw how androgenization of the female in fetal or neonatal life could alter patterns of gonadotrophin secretion in such a way as to make ovulatory cycles and behavioural oestrus impossible in adulthood. This is true of the rat, guinea pig, hamster and sheep, but it certainly does not apply to rhesus monkeys, marmoset monkeys or man. Androgens have been given to pregnant rhesus monkeys from the earliest stages of gestation, and although the female fetuses have shown extensive masculinization of their external genitalia at birth, and have subsequently been slow to attain puberty and have shown much rough-and-tumble play and male-type mounting behaviour, they have eventually shown oestrus and had normal, ovulatory menstrual cycles. Clinical evidence in man points to a similar picture; girls with congenital adrenal hyperplasia and extensive masculinization of their external genitalia can nevertheless ovulate and menstruate regularly after puberty, and we have already mentioned how true hermaphrodites with male external genitalia can do likewise.

It is an old adage that with increasing encephalization, mammals have become increasingly emancipated from the behavioural effects of their sex hormones, the absence of cyclical oestrous behaviour in women being a case in point. Our own sexuality may be determined by our culture, as much as by our hormones. However, even in man one has a feeling that androgens will ultimately make their presence felt, the best example being those 5α-reductase deficient men in the Dominican Republic who are born and reared as girls, but many of whom undergo a spontaneous change in their gender identity as their testes become fully functional at puberty.

Perhaps I have said enough to persuade you that sex hormones can produce almost as many dimorphisms in animal behaviour as they do in the phenotype. It will be a challenge for neuroanatomists to see if we can now uncover the anatomical basis for some of these spectacular organizational effects. Are masculinization and defeminization related to those changes in the medial preoptic nucleus that we described earlier? And are the long time-lags in the development of male-type behaviour the result of structural changes in the brain that must precede the behaviour?

Legal sex

Human sexual behaviour is such a very basic part of our lives, governing so many of our actions, that it should come as no surprise to find that the law has much to say on the subject. Legal attitudes to sex therefore provide a fascinating reflection of changing public attitudes.

Criminal law is obviously concerned with major sexual offences such as rape, and it is only slowly beginning to adopt a more tolerant attitude to the whole question of homosexuality. The recent movement for sexual equality is a commendable attempt to redress some of the excessive sexual dimorphisms in opportunity that have existed in our society in the past. Women were denied the vote, and were excluded from government, ostensibly because of men's subconscious fears that women were emotionally unstable at certain periods of the menstrual cycle, and that unforeseen pregnancies would constantly disrupt their lives. However, it would be a mistake to confuse the laudable objectives of equality of opportunity for the sexes, with the biological fact that the sexes do not possess equal abilities; men are physically stronger, 20 per cent heavier and about 13 cm taller than women, and hence they will always be better suited to certain manual tasks. The 'unisex' movement would make a great mistake if it tried to ignore the extent of human sexual dimorphism. As an interesting sidelight on this, the legal profession in England has recently reviewed the law relating to indecent exposure, and has concluded that it is an exclusively male offence, since the female has nothing indecent to expose.

The law is faced with a particular problem when it becomes necessary to decide whether somebody is male or female. This is of great significance in the laws of inheritance, and is becoming of increasing importance in international sporting events, such as the Olympic Games. Given the complexities of sex determination and differentiation, which factors should be taken into account in arriving at a decision?

The first instance in English law of a legal definition of sex was the case of Corbett vs. Corbett (otherwise Ashley) in 1970. A marriage was annulled when the 'wife', a man who had undergone a sex-change operation, was adjudged to be a male. In his summing-up, the judge said: 'The law should adopt, in the first place, the first three of the doctors' criteria, i.e. the chromosomal, gonadal and genital tests, and, if all three are congruent, determine the sex for the purpose of marriage accordingly, and ignore any operative intervention'. In cases where the three criteria are not congruent, 'greater weight should probably be given to the genital criteria than to the other two'. In many ways, this seems a sensible decision, since it is the appearance of the genitalia at birth that will have determined the assigned gender of rearing, and hence the gender identity of the individual. The chromosomal sex is certainly the least relevant factor; XY gonadal dysgenesis and XY testicular feminization cases should clearly be regarded as girls, and XX sex reversal cases should be regarded as boys. But problems can still arise. In the case of congenital adrenal hyperplasia girls born with extensive masculinization of the external genitalia, would it not be better to regard them as female, even if their behaviour is tomboyish, and surgically reconstruct their genitalia accordingly? And is it fair to deny men or women who have undergone sex-change operations the right to change their sex in the eyes of the law? We know enough about the complexities of human sexuality to realize how little we know, and surely the Jan Morrises of this world deserve the benefit of the doubt.

Sex determination is a simple event, in which maybe only a single gene is involved in deciding whether the indifferent gonad is to develop into an ovary or a testis. From that point onwards, the processes of sexual differentiation become increasingly complex as the gonad transmits its sexuality to the rest of the body by means of local and systemic hormones. Ultimately, almost all the somatic tissues of the body are caught up in sexuality. But we can only begin to appreciate the adaptive significance of these sex differences by understanding how Sexual Selection encouraged their development in the mating system characteristic of that species.

Suggested further reading
Effects of testosterone implants in pregnant ewes on their female offspring.
   I. J. Clarke, R. J. Scaramuzzi and R. V. Short. *Journal of Embryology and Experimental Morphology*, **36**, 87–99 (1976).
The sexual behaviour of prenatally androgenised ewes observed in the field.
   I. J. Clarke. *Journal of Reproduction and Fertility*, **49**, 311–15 (1977).
Sex determination in reptiles. J. J. Bull. *The Quarterly Review of Biology*, **55**, 3–20 (1980).
Sex determination and differentiation. R. V. Short. *British Medical Bulletin*, **35**, 121–7 (1979).
Differentiation of coital behavior in mammals: a comparative analysis.
   M. J. Baum. *Neuroscience and Behavioral Reviews*, **3**, 265–84 (1979).
Sexual behavior in red deer. R. V. Short. In *Animal Reproduction*, pp. 365–72.
   Ed. H. W. Hawk. BARC Symposium B. Allanheld, Osman, Montclair (1980).
Sex Chromosomes and Sex-linked Genes. S. Ohno. Monographs on Endocrinology 1. Springer-Verlag; Berlin (1967).
Major Sex-determining Genes. S. Ohno. Monographs on Endocrinology 11. Springer-Verlag; Berlin (1979).
Mechanisms of Sex Differentiation in Animals and Man. Ed. C. R. Austin and R. G. Edwards. Academic Press; London (1981).
Sexual Differentiation of the Brain. R. W. Goy and B. S. McEwen. MIT Press; London (1980).
Reproductive Endocrinology. Ed. S. S. C. Yen and R. B. Jaffe. Saunders; London (1978).

Sex, Hormones and Behaviour. Ciba Foundation Symposium 62. Excerpta Medica; Oxford (1979).

Sex, Evolution and Behaviour. M. Daly and M. Wilson. Duxbury; North Scituate, Mass. (1978).

Articles under the general title of 'Sexual dimorphism' by: F. Naftolin; J. W. Gordon and F. H. Ruddle; F. P. Haseltine and S. Ohno; J. O. Wilson; F. W. George and J. E. Griffin; C. W. Bardin and J. F. Catterall; N. J. MacLusky and F. Naftolin; B. S. McEwen; A. A. Ehrhardt and H. F. L. Meyer-Bahlburg; R. T. Rubin; J. M. Reinsich and R. F. Haskett. *Science*, **211**, 1263–324 (1981).

4

The fetus and birth

G. C. LIGGINS

Cut off from inquisitive eyes, as the fetus is, by the body wall of the mother and by the wall of the uterus, it is all too easily imagined as a sort of hothouse plant doing little more than grow during the weeks or months *in utero*, but ready to burst into full flower at birth. Even the greatest of fetal physiologists, Sir Joseph Barcroft, born 110 years ago, thought that the fetus lived in a sensory void, oblivious of its environment. But a moment's reflection is sufficient to reveal that birth represents no more than a transition from one environment to another and that most of the functions and activities to be observed in a newborn baby must have been present before birth. New techniques allowing study of the fetus in an undisturbed environment have convincingly confirmed that this is so. As a consequence there has been a tendency to replace the 'hothouse plant' concept by one that sees the fetus as a miniature adult, indulging in many of the activities of its extrauterine counterpart. This concept is little better, for on the one hand it minimizes the immaturity of many important organ systems at birth and on the other hand it ignores the many specialized functions peculiar to the fetus, upon which successful pregnancy depends. As this chapter proceeds, a picture will emerge of the fetus as an individual, dependent on the mother for the ingredients of growth and survival but nevertheless enjoying a remarkable degree of independence in the regulation of its development. And furthermore, by means of hormones secreted into the maternal circulation by the placenta or by the corpus luteum, the fetus can exert effects on the maternal physiology that enhance the mother's ability to meet the metabolic requirements of pregnancy. Finally, we shall see that not only may the fetus determine the timing of birth, but also, in anticipation of this event, initiate lactation.

Fetal growth

The growth and development of the fetus is determined in the main by the fetal genome. Superimposed upon the genetic regulation of growth are two opposing influences. On the one hand, fetal growth is constrained in various ways. For example, the supply of nutrients to the fetus is limited by the capacity of the mother and placenta for supply and transfer, respectively. On the other hand, a stimulus additional to the genetically determined drive to fetal growth and differentiation is provided by hormones and tissue growth factors. In early pregnancy, genetic control is dominant and determines relatively narrow limits of variability of patterns of fetal growth, whereas later in pregnancy constraints and stimuli assume increasing importance and give rise to greater variability.

Fetal diet. Carbohydrate is the main ingredient of the fetal diet. About half of the calories needed for growth and metabolism come from glucose, a quarter from lactate formed from glucose in the placenta, and the remaining quarter from amino acids. During maternal starvation more amino acids and less glucose are used. Fetuses have little or no capacity to synthesize glucose from fat or protein (gluconeogenesis) because the necessary enzymes, although present, are inactive at a low arterial oxygen partial pressure (pO_2). Gluconeogenesis begins promptly at birth when the arterial pO_2 rises.

In addition to calories, the fetus must have ample supplies of materials needed for the building of organs but which cannot be synthesized. They include essential amino acids, essential fatty acids, vitamins, minerals and trace elements. Many of these are selectively transferred from the mother to the fetus by active transport mechanisms that protect fetal needs at the expense of the mother in times of dietary deficiency.

Because of the importance of glucose in fetal growth it is worthwhile to consider its source and regulation in greater detail.

The source of glucose. The immediate source of glucose for the fetus is the maternal blood stream; more remotely, glucose comes from food consumed by the mother, from stores of glycogen, particularly in the liver, from fat depots and, during periods of starvation, from the breakdown of protein. The amount of glucose available to the fetus depends on its concentration in the maternal blood stream, which is maintained within rather narrow limits by a complex control system involving several endocrine organs. On the one hand, maternal glucose levels are prevented from falling by absorption of glucose from the gut and by the action of growth hormone, corticosteroids, catecholamines and glucagon which increase gluconeogenesis and liberate glucose from glycogen stores. On the other hand, an increase in glucose concentration above normal levels is prevented by insulin, which increases both the breakdown of glucose in muscles and its passage into glycogen or fat stores. This system serves the adult well and ensures adequate supplies of glucose in a wide variety of stressful situations. Yet these safeguards do not always seem sufficient to satisfy the demands of reproduction, and certain pregnancy hormones may further modify maternal glucose control in a way that increases availability of glucose to the fetus. In many species, during the first half or more of pregnancy there is a substantial increase in maternal body weight due mostly to the deposition of subcutaneous fat. This is an effect of progesterone which not only increases appetite but also diverts glucose into fat synthesis. Later in pregnancy when the metabolic requirements of the fetus are at their peak, or during periods of starvation, these extra fat stores are used to maintain a constant supply of glucose to the fetus. Pregnant primates have an additional hormone called placental lactogen, which is secreted in large amounts into the maternal circulation. Placental lactogen is closely related chemically to growth hormone, and although of low biological activity it shares with growth hormone an effect in elevating blood glucose levels.

The passage of glucose from mother to fetus is encouraged by 'facilitated diffusion' which maintains a constant relationship of maternal and fetal glucose concentrations, the maternal level being higher by a constant amount. In other words, the concentration of glucose in the fetus is determined by the mother. The amount of glucose reaching the placenta is the product of the concentration in the blood and the volume of blood passing through the placenta. Deficiencies in either or both can occur and may lead to retardation of fetal growth. In many animals, lower blood glucose levels resulting from inadequate nutrition are usually at fault, but in the better-fed human subject, disorders of blood supply are more common.

Fetal regulation of glucose utilization. One wonders whether the rate of fetal growth is constantly restrained by availability of glucose, or alternatively whether there is normally more than enough at all times, so that growth is restrained by the mechanisms that control glucose utilization. We will try to answer this question in a succeeding paragraph. But first we must consider the ways in which the fetus differs from the adult in its glucose balance.

As we have already seen, the levels of glucose in the fetus mirror those in the mother. Thus there is neither the need nor the opportunity for the fetus to regulate its own blood glucose concentrations. The mechanisms for doing so do develop in the fetus, but are largely dormant until birth, when the supply of glucose from the placenta ends abruptly. Nevertheless, two important processes are active from an early stage. The first of these, which we will discuss in more detail later in the chapter, is concerned with the storage of glucose as glycogen or fat to provide for the metabolic needs of the newborn baby until feeding begins. The second is the control of the rate at which glucose is used by the growing tissues, and this is probably owing to the action of insulin secreted by the fetal pancreas. When insulin is administered to the fetuses of experimental animals such as rats and sheep, the rate of growth is considerably accelerated despite low fetal blood sugar levels (hypoglycaemia), and there is an increase in the size of glycogen-storing organs such as the liver and heart, as well as increased deposition of fat. A similar phenomenon can be observed in the overweight newborn babies of diabetic mothers (Fig. 4.1). In this disorder, lack of maternal insulin secretion causes elevation of blood glucose levels. Fetal blood levels rise with those of the mother and stimulate the release of insulin from the fetal pancreas, which leads in turn to increased growth and fat storage. Thus we are led to the conclusion that the rate of glucose utilization rather than glucose availability is the more important factor regulating growth under conditions of adequate nutrition.

Fig. 4.1. Newborn babies who illustrate the effects of too little or too much insulin during fetal life. Baby *A* was himself diabetic at birth due to defective secretion of insulin by his pancreas. He shows signs of intrauterine starvation because of his limited ability to use glucose. On the other hand, Baby *B* is overweight and fat because of over-secretion of insulin by her pancreas, leading to over-use of glucose. This was because her mother was diabetic, and so her blood glucose levels were elevated during pregnancy.

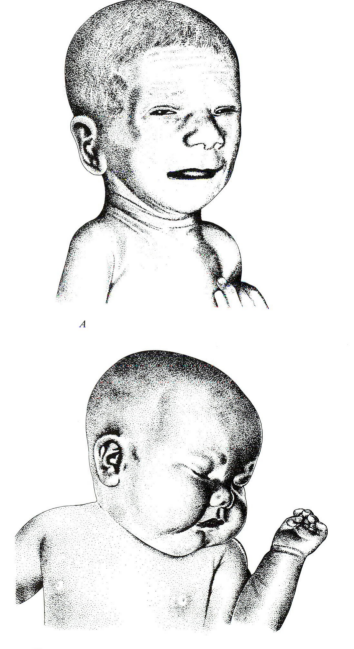

A

B

Somatomedins. The somatomedins are a family of peptides, so named because they are believed to mediate the action of growth hormone (somatotrophin) on postnatal growth. They are structurally related to proinsulin (the larger protein from which insulin is made) and have insulin-like metabolic actions. Two of the somatomedins, insulin-like growth factor I (IGF 1) and IGF 2, may be involved in fetal growth because of their relationship to insulin. Fetal cells contain separate receptors for insulin, IGF 1 and IGF 2, but insulin reacts with IGF receptors and IGF reacts with insulin receptors. Insulin probably expresses some of its effects on growth through somatomedin receptors. The main source of somatomedins is the liver but little is known about how their synthesis is regulated. Present evidence favours insulin as the more important hormone in overall fetal growth and places somatomedins in a supporting role, particularly in stimulating mitosis at specific sites.

Cell growth

So far we have been concerned with growth of the whole body or of whole organs. Clearly, however, change in size of the body or of a particular organ is the result of increments in cell number and cell size. Cell multiplication is the most fundamental activity in fetal tissues, and when associated with the equally remarkable process of differentiation it forms the basis of development. Starting with a single cell, the fertilized egg, complete development of the human fetus to term involves 42 cell divisions, yet only a further five divisions are needed to attain adulthood.

The cells of the developing organism have a strong inherent capacity to multiply and to differentiate. Retarded growth due to a reduction in cell numbers is uncommon, occurring in association with congenital malformations, severe fetal viral infections and exposure to antimitotic drugs. Control of cell division and cell differentiation is exerted mainly within the tissues themselves by local 'organizers' synthesized within the cells in response to genetic information contained in their chromosomes. Cells must have a precise mechanism for counting replications but its nature in mammalian tissue is unknown. In micro-organisms, each cell replication may alter a nucleotide in successive codons of a special segment of the DNA strand. After a certain number of replications, a codon is reached that inhibits further replication when modified. Retarded fetal growth is more commonly the result of a reduction in cell size, due either to lack of calories or an inability to use calories effectively because of deficiency of hormones such as insulin, or substrates such as essential amino acids or fatty acids.

Fetal respiration

Respiration, by which is meant oxidation of foodstuffs (mainly glucose in the fetus) with the liberation of energy and carbon dioxide, is essential for the support of life and the processes of growth. Factors concerned with the supply of glucose to the cell have already been discussed and it is necessary now to turn our attention to the gases, oxygen and carbon dioxide.

During fetal life, the placenta, not the lung, is the organ of gaseous exchange, and although anatomically these two organs appear so different their respiratory functions have much in common. In essence, each consists of membranes that are permeable to oxygen and carbon dioxide and through which the gases diffuse readily along a concentration gradient. In lungs, the gradient for oxygen is from the partial pressure of oxygen in the terminal air sacs ($\simeq 100$ mmHg) to that in arterial blood ($\simeq 90$ mmHg). In the placenta, however, the gradient is from the partial pressure of oxygen in the maternal arterial blood ($\simeq 90$ mmHg) to that in fetal umbilical venous blood ($\simeq 25$ mmHg). For various reasons the partial pressure of oxygen in the fetal blood cannot approach closely that in the maternal blood. For one thing, the placenta, which consumes considerable quantities of oxygen, lies between maternal and fetal circulations. Furthermore, the relationship of the fetal blood vessels to the maternal vessels is not a perfect counter-current system, and thus an equilibrium will be reached in the fetal blood which lies somewhere between the partial pressures of oxygen entering and leaving the maternal side of the placenta (Fig. 4.2). In fact the pO_2 in fetal blood leaving the placenta is less than one-third of maternal arterial pO_2. Nevertheless the universal success of pregnancy suggests that the fetus suffers no disadvantage from a low arterial pO_2. The fetus should not be thought of as chronically short of oxygen, nor equated, as some of the older physiologists

Fig. 4.2. Blood-gas exchange across the human placenta. The figures for partial pressure of oxygen and carbon dioxide are expressed in millimetres of mercury.

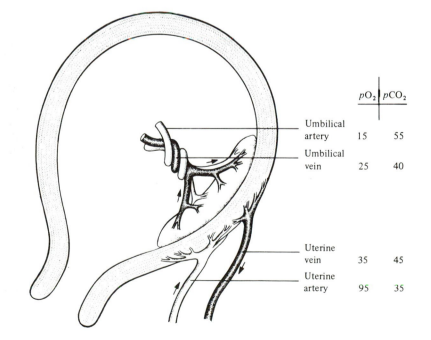

	pO_2	pCO_2
Umbilical artery	15	55
Umbilical vein	25	40
Uterine vein	35	45
Uterine artery	95	35

did, with a man on the top of Mount Everest. The fetus is quite readily able to extract from its blood the volume of oxygen necessary for respiration, a volume in the human and sheep fetus near term of approximately 15 ml/min. This is possible mainly because of a high cardiac output and a large circulating blood volume (120 ml/kg compared to 70 ml/kg in adults), ensuring a large blood flow both to the placenta and to fetal tissues. In addition, late in gestation in some species, the oxygen dissociation curve of intact red cells is shifted to the left of that of the mother, and this allows more oxygen to be carried at the same pO_2.

The fetus appears to lack control over the amount of maternal blood perfusing the placenta. Nevertheless the pregnancy hormones, oestrogen and progesterone, have ensured that the blood supply is adequate by promoting growth of the blood vessels and maintaining their dilated state throughout pregnancy. Towards term in man, at least 10 per cent of maternal cardiac output (which is itself increased by up to 25 per cent compared to the non-pregnant state) reaches the placenta; thus in normal circumstances, the oxygen supply to the fetus is not only adequate to provide for the requirements of growth, but also has sufficient reserve to permit fetal survival when uterine blood flow is intermittently interrupted by the uterine contractions of labour. But in some circumstances, a deficient maternal blood supply may lead to retardation of growth or even to fetal death. In man, for example, the maternal blood vessels may be damaged by complications of pregnancy such as hypertension.

Placental growth

A discussion of fetal growth would be incomplete without including the placenta, for not only is the placenta largely composed of fetal tissue, but also there is a strong correlation between the size of the fetus and the size of its placenta. This association may or may not represent a causal relationship. Those factors affecting fetal growth could be expected also to affect placental growth. On the other hand, placental size might limit fetal growth by limiting either the transfer of nutrients or the production of hormones. Fetal growth retardation can be induced in experimental animals by removing portions of the placenta, but generally more than half the placenta must be excised to be effective and the remaining placenta has a considerable capacity for compensatory growth. This suggests that placental growth is adapted to the needs of the fetus, but how this might be accomplished is unknown.

Fetal functions

A variety of features of the intrauterine environment relieve fetal organs of the need to perform certain functions. For example, the placenta substitutes both for the lungs as organs of gas exchange and for the kidneys as organs of excretion. The weightless state of the fetus removes the need for muscular strength and the isothermal environment means that calories are not expended in maintaining body temperature. The gastro-intestinal tract does not exercise a digestive function and the liver receives no products of digestion in the portal vein. These examples of lack of function might suggest fetal dormancy, but as we shall see, all these body systems must reach a degree of development before birth sufficient to sustain life immediately after birth. This is well illustrated in species such as sheep in which the newborn lamb is relatively mature; it stands, breathes and feeds within minutes of birth, thus revealing the potential attained as a fetus. In the remaining paragraphs we will discuss, system by system, the special features of function distinguishing fetal from extrauterine life.

Cardio-vascular system

The circulatory system of the fetus is essentially that of the adult adapted to the special circumstance that the placenta functions in place of the lungs as the organ of respiration. This may sound like a relatively minor difference, but the adaptation is a feat that any hydraulics engineer would be justifiably proud to accomplish. It entails the diversion of almost the entire output of the right ventricle of the heart away from the lungs and into the umbilical circulation; and what is more, the diversion is achieved in a way that allows almost instantaneous conversion to the adult type of circulation as the first breath is taken. The anatomical solution to the problem is straightforward – the pulmonary artery and the aorta develop with a large connection between them so that blood heading towards the lungs is diverted directly into the aorta and thence to the placental circulation.

The functional solution to the problem of redirecting blood to the lungs when the first breath is taken is also straightforward, though remarkably ingenious. The diversionary passage, the ductus arteriosus, has a muscular wall that is sensitive to the partial pressure of oxygen in the blood and responds to an increase in oxygen tension by contracting and closing. Thus, when arterial oxygen tensions rise with the onset of breathing, the ductus will tend to close (see Fig. 4.3). Before birth there is little blood going to the lungs and therefore little returning from the lungs to enter the left atrium of the heart. In the adult system this would mean that almost all

Fig. 4.3. The fetal circulation. This diagram differs from that for an adult in showing a placental circulation, a ductus arteriosus and a foramen ovale.

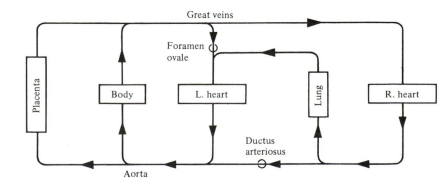

of the venous return to the heart, including blood from the placenta, would enter the right atrium. In the fetus the difficulty is overcome by the presence of a hole in the septum between the right and left atria allowing half of the blood entering the right atrium to pass to the left atrium. The hole, the foramen ovale, is cunningly designed as a flap valve that not only diverts the better-oxygenated blood of the inferior vena cava into the left atrium, and thence to the brain, but also shuts at birth when the blood pressure in the left side of the heart rises to a level above that of the right side. Circulatory adaptation of the fetus thus depends on two diversionary passages, one a sphincter responsive to oxygen tension, the other a flap valve responsive to pressure, both of which close at birth when breathing begins.

Gastro-intestinal system

Gastro-intestinal function in the human fetus has been extensively investigated by means of intra-amniotic injections, either of opaque media demonstrable by X-rays or of inert chemicals which can be transferred to the mother and excreted in her urine. In the latter part of pregnancy the human fetus swallows about 500 ml amniotic fluid daily. When opaque media are included in the swallowed fluid it can be shown to pass rapidly into the stomach and onwards into the small bowel, reaching within a few hours the large bowel where it remains throughout the rest of pregnancy. Most of the swallowed material is water which is rapidly absorbed in the small bowel. Other materials of small molecular size such as electrolytes, glucose, urea and steroid hormones are also absorbed. Large molecules, cell debris and the sebaceous (fatty) material secreted by the glands of the fetal skin are not absorbed, but accumulate in the large bowel, where, together with cells shed from the intestinal wall and bile pigments, they form the green faecal material known as meconium. Under normal circumstances defaecation *in utero* does not occur; but following acute oxygen deprivation, the large bowel may contract and the rectal sphincter relax, allowing the meconium to be discharged into the amniotic fluid.

We are not clear what benefit derives from the swallowing of such large amounts of fluid, but undoubtedly the practice helps to prepare the intestines for their digestive functions after birth. The swallowing of amniotic fluid may also contribute to fetal nutrition: approximately a gram of protein is swallowed daily and most of this is probably absorbed. Fetuses malformed by the failure of the oesophagus to develop as an open tube are of lower birthweight than normal babies, suggesting that the swallowed fluid has nutritive value. Probably the most important function of swallowing is concerned with control of amniotic fluid volume. Excessive amniotic fluid is a usual accompaniment of disorders of fetal development that impair ability to swallow normally. This topic will be discussed further in the section on renal function.

The mouth of the newborn primate is much more than an orifice through which food is ingested; perhaps because of the ability to hold objects and to put them against the lips, the mouth has become a sensory organ of some importance in early postnatal life. It is not surprising, therefore, that some evidence of this aspect of the function of the mouth has been seen in the human fetus – X-rays not uncommonly reveal a fetus apparently sucking a thumb! The sensation of taste also seems to be present *in utero*. Experiments in which the rate of swallowing has been measured by determining the rate of disappearance from the amniotic fluid of red cells labelled with radioactive chromium have shown that the addition of saccharine to the amniotic fluid increases the rate of swallowing, whereas distasteful materials such as opaque media cause almost complete cessation of swallowing. The taste buds of fetal sheep are mature by the beginning of the last third of pregnancy and respond like those of an adult to salt and other tasty stimuli. The sucking reflex on which postnatal survival is so dependent is amongst the earliest of the co-ordinated reflex activities to appear in the developing fetus. In some fetal animals, strong sucking can be elicited before the middle of pregnancy by placing a fingertip between the lips.

Renal system

Throughout pregnancy the placenta, not the kidney, is the organ of excretion. Waste products are transferred into the maternal circulation to be voided into the maternal urine. The fetal kidney might therefore seem to have no important function. Observations on fetuses, both animal and human, in which kidneys have failed to develop or in which there is complete obstruction of the urinary tract appear to bear this out; the fetuses survive to term and are born alive. However, such pregnancies are not entirely normal, suggesting that the fetal kidneys may have functions other than those concerned with excretion. Growth is retarded in fetuses whose kidneys have failed to develop (renal agenesis) and also in fetal lambs from which the kidneys have been removed earlier in pregnancy. Paradoxically, growth hormone levels in the blood of the fetal lambs become abnormally elevated after removal of the kidneys. One suggested function of the kidney is to convert the large protein molecule of growth hormone to a smaller, biologically active molecule.

Another anomaly observed in pregnancy when the fetus lacks kidneys is the virtual absence of amniotic fluid. The mature normal human fetus may excrete 450 ml of urine daily, a volume that would seem a significant contribution to a total amniotic fluid volume of about 1500 ml. But the hourly turnover of water in the amniotic fluid is 300–600 ml, so that the contribution of urine is really very small. Fetal urine has been suggested as an important source of amniotic fluid protein, which provides the osmotic pressure necessary to retain water within the amniotic sac. Loss of protein from amniotic fluid, as we are already aware, occurs by fetal swallowing. The total mass of amniotic fluid could be regulated by the balance between fetal swallowing and fetal micturition. Aberrations of amniotic fluid volume associated either with defective fetal swallowing or defective micturition certainly are consistent with this. However, the amount of protein in fetal urine is too little to account for anything but a small part of the total protein turnover in amniotic fluid, and although total protein mass is likely to be a major factor regulating amniotic fluid volume, neither the source of the protein nor the mechanisms controlling its entry into amniotic fluid are known.

In those mammals with a persistent allantoic sac (this does not include

Fig. 4.4. An ultrasound image of a human fetus at 24 weeks of pregnancy. Real-time scanning techniques allow continuous observations of fetal activities such as limb movements, thumb sucking, urination and 'breathing'. (Courtesy of Professor Stuart Campbell.)

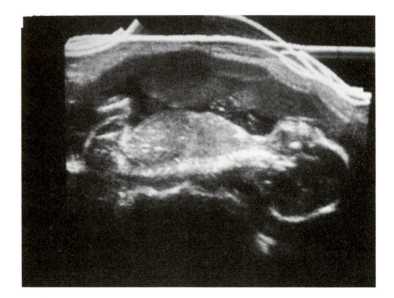

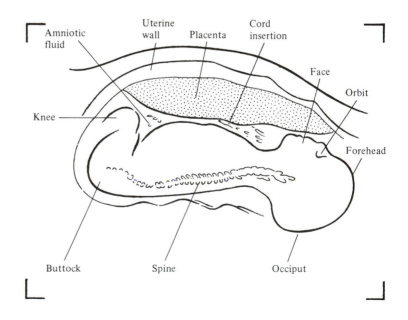

humans), the fetus can void urine either through the urachus to the allantoic sac or through the urethra to the amniotic sac. In early pregnancy most of the urine probably goes to the allantoic sac, but as pregnancy progresses more is voided into the amniotic sac.

Respiratory system

Throughout at least the latter half of pregnancy, fetuses of all species spend a third or more of each day in making breathing movements. The movements are rapid (1–4 Hz) and irregular in amplitude and frequency, and generate negative pressures of 25 mmHg or more in the chest. The remarkable feature of the movements is that they occur in episodes of up to 30 minutes, only during rapid-eye-movement (REM or 'dreaming') sleep. During wakefulness and quiet sleep, breathing movements are absent. Unlike extrauterine breathing in which the chest wall is active, fetal breathing movements involve only the diaphragm, the muscles of the chest wall being quiescent. Of course, there is no air to be inhaled and exhaled, only fluid; and the inertial mass of fluid is so great and the breath so short in duration that less than 1 ml of fluid moves with each breath through the larynx of mature fetal sheep. Human fetal breathing movements can often be seen simply by observing the abdominal wall of the mother, and they are readily studied with ultrasound devices (Fig. 4.4). Contrary to older views that fetuses breathed only when asphyxiated, we now know that they breathe normally only when healthy. Recordings of human fetal breathing are used clinically to recognize fetuses in danger of dying from a complication of pregnancy.

The function of fetal breathing movements is two-fold. First, they allow development by 'practising' of the complex neuromuscular activities needed to ensure that the first breath will expand the lungs with air. Second, they distend the lungs with fluid and stimulate their growth. When fetal breathing is hampered by congenital abnormalities of the nervous system or diaphragm, the lungs will be small and incapable of sustaining extrauterine life.

Endocrine system

The endocrine organs are functional by the end of the first quarter of pregnancy in the human fetus and thereafter the fetus is self-sufficient in its hormonal requirements. In general, the fetal glands secrete the same hormones with the same functions as those of the adult. But this information conveys nothing of the complexity and fascination of fetal endocrinology, for not only do the fetal organs subserve a number of unique functions vital to normal development, but there are also complicated interrelationships between mother and fetus.

A detailed description of one of the most interesting of the fetal endocrine systems, the feto-placental unit, will be found in Book 3. This unit, found in man and probably some other mammals, is so named because the complete elaboration of certain hormones (particularly oestro-gens) secreted by the placenta needs the combined efforts of both the placenta and the fetal organs. Enzymes concerned in the biosynthesis of the hormones are distributed amongst the various tissues so that no one organ is capable of complete synthesis. Of the many and varied special problems in fetal endocrinology apart from the feto-placental unit, it is only possible within the scope of this chapter to cite one or two examples relating to each of the endocrine organs. But first, let us state some principles that describe general hormonal relationships between the mother and the fetus.

1. The fetal endocrine system profoundly influences the mother. The hormones secreted by the feto-placental unit into the maternal circulation are responsible for most of the physiological changes that occur in the mother during pregnancy, changes involving every system of her body. For example, alterations in the function of the maternal central nervous system are apparent in altered behavioural patterns; her heart output is increased and her peripheral blood vessels dilated; in the kidneys there is an increase in glomerular filtration rate; the blood levels and secretion rates of several maternal hormones are increased; the oestrous or menstrual cycle is suppressed; glucose, fat and protein metabolism is altered; ligaments relax, and the mammary glands develop. In mammals that are not dependent on the corpus luteum for maintenance of pregnancy (these include man, horse, sheep and guinea pig) the changes are induced directly by placental hormones, while in those species that require a functional corpus luteum throughout the whole of pregnancy (such as rabbit, pig, rat and goat) the influence of the conceptus is exerted in part by hormones regulating the function of the corpus luteum.

2. The maternal endocrine system does not directly influence the fetus except in disease. The protein and polypeptide hormones of the pituitary and pancreas are completely excluded from the fetus by their inability to pass through the placenta. Hormones of smaller molecular size such as the adrenal corticosteroids can pass readily through the placenta of some species (as in man) but less readily in others (like the sheep); but even when passage does occur, much of the active hormone may be metabolized to biologically inactive material by specific enzymes in the placenta. In those species that have been investigated, thyroxin has been found to pass through the placenta in insufficient amounts to provide for the needs of mature fetuses. Alternatively, a maternal endocrine organ may influence its counterpart in the fetus by indirect means; insulin is a good example of this (see below).

3. The fetal endocrine system is autonomous. As a consequence of the exclusion from the fetus of maternal hormones, the fetal system is self-regulating and meets the special needs of the fetus.

4. The fetal endocrine system has specialized functions not present in adult life. At certain critical phases of development fetal hormones have special functions never again required of them. Examples, such as sex determination, preparation for birth and initiation of labour, are described in later sections of this chapter.

The hypothalamus. The maternal hypothalamus is described in detail in Book 3. Here we should remember that the hypothalamus is active during fetal life and that it co-ordinates the activities in the other endocrine organs in the same way that it does in the adult. In one way, however, it is strikingly different from the adult hypothalamus – in the course of maturation the function of the fetal hypothalamus can be permanently modified by brief exposure to certain hormonal stimuli. This is discussed more fully in the preceding chapter. Other intrauterine factors may possibly leave imprints on the hypothalamus that could permanently modify behaviour. This is one good reason why drugs, even those that have no teratogenic effects, should be used with caution in pregnant women.

The pituitary gland. The anterior pituitary gland is functional throughout most of fetal life and it secretes the same trophic hormones that are secreted by the adult gland. In general, the trophic hormones have functions similar to those of the adult, stimulating both secretion and growth in the target organs. Fetal hypophysectomy prevents normal development of the thyroid, adrenals and testes, but the pancreas and ovaries are little affected.

When the fetal pituitary is separated from the hypothalamus by dividing the pituitary stalk, the signs of defective pituitary function are much less distinct than after hypophysectomy, suggesting that the fetal pituitary is less dependent than the adult pituitary on releasing hormones from the hypothalamus. Nevertheless, the usual feedback mechanisms act on the hypothalamus to regulate trophic hormone release from the pituitary. Administration to the fetus of cortisol, the major secretory product of the adrenal cortex, causes adrenal atrophy by inhibiting the release of corticotrophin (ACTH). This effect of cortisol is shown particularly clearly in human pregnancy when corticosteroids are administered to the mother. Passage of the corticosteroid across the placenta causes a sharp reduction in adrenal secretion of the precursor material used by the placenta in the biosynthesis of oestrogens. Consequently there is a steep fall in the excretion of oestrogens in the maternal urine.

The thyroid gland. Thyroxin is particularly important in the normal development of brain, bones, hair and wool. Newborn animals suffering from the effects of hypothyroidism *in utero* may be of normal size, but show behavioural retardation, have a bone age that has lagged far behind the chronological age, and are deficient in body hair. These changes are present, even though maternal levels of thyroxin are normal, because the restricted passage of thyroxin across the placenta allows insufficient entry of maternal thyroxin to meet fetal needs. A very complicated feto-maternal endocrine relationship may develop in the course of the human thyroid disease, thyrotoxicosis. The manifestations of this disorder are the result of excess secretion of thyroxin from a gland that is under constant stimulation by an abnormal thyroid antibody. Although thyroxin crosses the placenta with difficulty, the antibody crosses more readily, causing excess activity in the fetal thyroid and hence fetal thyrotoxicosis. Furthermore, the thyroid-blocking drugs used in the treatment of maternal thyrotoxicosis also cross the placenta and if used injudiciously may suppress function in the fetal gland. Not only may the fetus then show evidence of hypothyroidism, but also it may develop a goitre because low thyroxin levels lead to hypersecretion of thyrotrophin from the fetal pituitary.

The fetal pancreas. Of the various hormones concerned in the regulation of growth, insulin from the fetal pancreas is probably the most significant because it regulates the rate of glucose utilization which, as we have already seen, is the source of energy for fetal growth. It is intriguing to wonder how the release of insulin from the fetal pancreas is regulated. In the normal adult the insulin response to taking food is determined in two ways: first, a hormone that stimulates insulin secretion is released from the lining of the intestine when glucose in the ingested food comes into contact with it. Consequently, a rise in the concentration of insulin in the blood stream precedes the rise in glucose. Second, glucose has a direct stimulatory effect on the pancreas. In the fetus, insufficient glucose is swallowed to activate

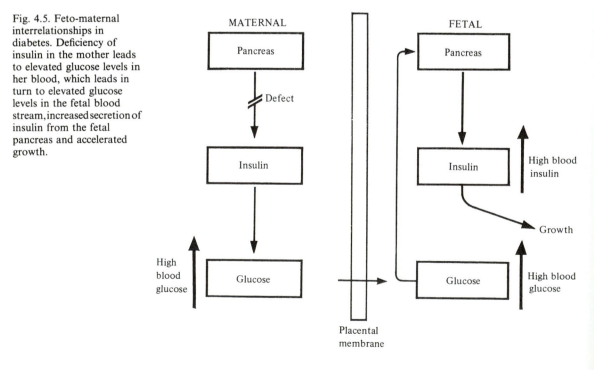

Fig. 4.5. Feto-maternal interrelationships in diabetes. Deficiency of insulin in the mother leads to elevated glucose levels in her blood, which leads in turn to elevated glucose levels in the fetal blood stream, increased secretion of insulin from the fetal pancreas and accelerated growth.

the intestinal hormone system. Furthermore, the fetal pancreas responds to elevations in blood glucose levels with only a small rise in insulin secretion. Thus, insulin secretion in the fetus seems likely to be continuous and relatively unvarying. Nevertheless, under abnormal circumstances the fetal insulin secretion can be abnormally high. In diabetic women who have elevated blood glucose levels over long periods of time, excessive growth of the fetal pancreas occurs, no doubt in response to a sustained elevation of fetal blood glucose levels (Fig. 4.5). The consequences of increased insulin secretion from the fetal pancreas on the rate of growth have already been discussed.

The fetal gonads. The fetal ovary in most mammals is small and shows little evidence of activity. In the preceding chapter on sex determination, the passive role played by the ovary has been described. The extraordinary phenomenon of ovulation that occurs in the fetal giraffe is likely to be of an incidental nature, serving no useful purpose. The fetal testis, on the other hand, is a highly active endocrine organ whose function is essential for normal sexual differentiation in the male. The development and function of the testis is regulated by the hypothalamus and pituitary, as can be readily demonstrated by observing the effects of fetal hypophysectomy early in pregnancy. The interstitial cells of the testis are also stimulated by human chorionic gonadotrophin (hCG) and it is tempting to link the peak of hCG secretion at 60–70 days with the time of sexual differentiation.

The fetal adrenal glands. No particularly important function has been attributed to the medulla of the fetal adrenal. Like the adult adrenal medulla it secretes both adrenalin and noradrenalin. It differs from the adult gland in containing a higher concentration of noradrenalin than

Fig. 4.6. A diagram to illustrate the multiple functions of the adrenal gland in the control of events leading up to birth.

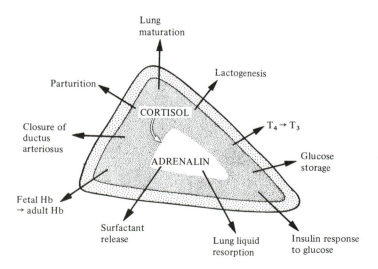

adrenalin but the proportion of adrenalin increases towards term. The large amounts of adrenalin secreted during labour may have an important function in promoting maturation of the lung. In addition, noradrenalin and adrenalin help the fetus to tolerate birth asphyxia by contributing to the maintenance of a direct response to asphyxia as well as the adult-type response mediated by the splanchnic nerves. The release of catecholamines from the adrenal medulla probably aids survival during birth asphyxia by contributing to the maintenance of the circulation and by mobilizing glucose from body stores.

The cortex of the adrenal is the most versatile of the fetal endocrine glands (Fig. 4.6). The large number of specialized functions attributed to it fall into two groups. The first group includes actions that have in common timing of developmental events; all of these depend upon the ability of cortisol to induce activity in specific enzyme systems. Such cortisol-inducible enzymes have been identified in a variety of fetal organs including brain, liver, eye, intestine, adrenal medulla, placenta and lung. The developmental consequences of 'switching-on' an enzyme system may be profound and some of the physiological events that follow are described below in the sections entitled 'Preparations for birth' and 'Control of parturition'.

The second group of functions of the adrenal cortex depends on the development in certain species of a specialized area of the cortex known as the fetal zone. This reaches its most highly developed form in the human fetus where its bulk is so great that the total weight of fetal adrenal tissue is equal to that of the adrenal from a child of 10–12 years. After birth the fetal zone regresses quite rapidly and by 2–3 months of age has disappeared. Biochemically, the zone is distinguished by its inability to synthesize cortisol. Instead, large quantities of dehydroepiandrosterone sulphate, the precursor of oestrogen, are secreted into the fetal circulation. Further details of this aspect of adrenal function can be found in Book 3 in the section describing the feto-placental unit.

Immunological system

There are many remarkable mechanisms involved in successful reproduction but none more so than the means by which the conceptus is able to avoid immune rejection by the mother. The mammalian fetus is attached to the uterine wall in a way that resembles a graft, and yet genetic differences exist between the mother and her fetus. Half the genetic make-up of the fetus is derived from the mother, and half from the father. The maternal component poses no problem of incompatibility, but the paternal component is foreign to the mother and might be expected to lead to rejection. But even embryos from transplanted eggs, in which both the maternal and paternal components are foreign to the donor mother, can implant and develop normally. How this tolerance is achieved is an unresolved mystery.

(These and other aspects of the immunology of the fetus are discussed in Chapter 2 of this book and in Chapter 6 of Book 4.)

Nervous system

In the young animal we are used to the idea that the development of brain function for motor activities, for perception, and for behavioural patterns is to a great extent dependent upon, and modified by, environmental influences; genetic pre-programming seems less in evidence. This is not to say, of course, that such pre-programming is absent. To take a simple example, a group of young animals reared in isolation from their elders will develop the general patterns of social behaviour characteristic of the species. In the fetus, on the contrary, the development of brain function follows a course that is mainly genetically determined and is largely independent of environment and of peripheral events within the fetus. In mammals with long gestation lengths the sensory organs of the fetus are highly developed and are capable of response to the usual stimuli. Although the intrauterine environment is by no means the sensory void pictured by physiologists of old, neither do the scope or intensity of stimuli approach that of the extrauterine environment. The usual intensity of sound, for instance, within the human uterus is approximately 50 decibels (comparable to sound intensity in a quiet office) reaching a maximum of 95 decibels with maternal heart beats, and the intensity of light (unless the mother happens to be sunbathing in a bikini) is similar to that of a darkened lecture room. Thermal stimuli are almost completely absent, the intrauterine temperature being maintained within very narrow limits at about 0.5 °C above that of the mother. Tactile stimuli are minimized by the cushioning effects of the amniotic fluid, and the state of near-weightlessness favours a low input of proprioceptive information. Nevertheless, the fetus can readily be shown to be responsive to unaccustomed levels of any of these stimuli. A loud noise close to the uterus, a flash of high-intensity light, the touch of a needle tip on the skin, a rapid change in the position of the mother, or the injection of a cold fluid into the amniotic sac, all may evoke a quick and vigorous reaction in the form of fetal movement and acceleration of heart rate.

Whether the relatively unvarying and unexciting information about the immediate environment that reaches the fetal brain through the nervous system has significant effects on brain development is uncertain. A greater impact may be made by chemical agents reaching the fetal brain from the maternal circulation. The severity of withdrawal symptoms of a newborn infant addicted to heroin while *in utero* may be life-threatening. Less dramatic effects may result from nicotine or from abnormal amounts of adrenalin in a chronically stressed mother. Nevertheless, there is little evidence that would support those physiologists who would have us believe that maternal behaviour can have effects on brain development that play an important part in moulding the personality of the child and adult. Imprinting can and does occur *in utero* (such as the effects of androgen on the hypothalamus) but is the result of genetically controlled events within the fetus rather than of random events in the environment.

To the mother, whether human or animal, fetal movements are her only certain evidence of her pregnant state. They start almost as soon as the limbs are formed, but until the fetus becomes bigger, movements do not make their presence felt. That they can be felt by the human mother as early as the 14th week is an indication of their vigour. The purpose served by fetal movements is not altogether clear. Part of their function is concerned with the development of muscle action by 'exercise' but fetal muscles develop, at least anatomically, in the absence of nervous connections. Indeed, in species such as the rat, the muscles are well formed before nerves have reached them, and motor end-plates begin to develop only about 4 days before birth. In mammals that have longer gestation periods, and that are born in a more mature state than the rat, motor end-plates appear during the first half of pregnancy, allowing a greater opportunity for the motor nerves to dominate further maturation of the muscles.

If a fetal lamb still attached to the mother by the umbilical cord is observed in a warm water-bath, three states of movement can be distinguished. In the first state, called 'quiet sleep', there is little or no spontaneous movement and the response to stimuli is sluggish. In the second state, termed 'wakefulness', the lamb opens its eyes, seems aware of its surroundings, responds briskly to stimuli, and there is much spontaneous movement of the limbs and trunk. The movements probably have little to do with what is happening in the brain and are probably the result of spinal reflex activity. Indeed, headless monsters are remarkable for their abnormally high motor activity, suggesting that the brain may have an inhibitory rather than excitatory influence. The third state, 'rapid-eye-movement sleep' (REM sleep, active sleep) is characterized by twitching movements of the legs, trunk and facial muscles and by rapid horizontal movements of the eyes. Stimuli usually have no effect other than to cause cessation of the twitching movements.

REM sleep is an intriguing phenomenon that has so far defied a satisfactory explanation. In adult individuals deprived of sleep, restoration of the deficit of REM sleep gets priority over other sleep states, suggesting that it has functions of some importance. One hypothesis suggests that during REM sleep the brain computer 'reprogrammes' its more complex perceptive and motor functions. This idea has something to commend it in the fetal lamb which spends at least a third of the day in this form of sleep; three of the most highly co-ordinated motor activities that occur after birth – eye movements, breathing movements and movements of the mouth – are present more or less continuously through REM sleep.

The development of function in the fetal brain is the darkest corner in fetal physiology. New tools such as fetal electroencephalography are now beginning to allow study of the undisturbed fetus *in utero* and can be expected to advance our knowledge rapidly.

Preparations for birth

At birth, separation from the placenta suddenly deprives the newborn animal of the supplies of oxygen and glucose to which it has been accustomed. Unless an alternative supply is found rapidly, death will ensue. In normal circumstances breathing starts within seconds of birth, and is fully established after a few minutes. Glucose is quickly released from carbohydrate reserves, particularly from glycogen stored in the liver and heart muscle. This source of glucose is exhausted within a few hours and fat then becomes the main metabolic fuel supplying both fatty acids and glycerol for oxidation in the tissues. Even the fat stores can sustain life for only a limited period, but this is usually more than sufficient to bridge the variable gap until suckling starts. Successful transition from intrauterine to extrauterine existence is thus clearly dependent on events that have preceded birth – functional maturation of the lungs, laying down of carbohydrate and fat reserves, and the onset of lactation. The regulation of many of the final preparations for birth resides in the fetal adrenal glands which become increasingly active as term approaches.

Maturation of lung function

Given reasonable anatomical development, the lungs can be maintained in an air-distended state after birth provided that an adequate concentration of a surface-active material called 'surfactant' is present in the air sacs (alveoli). This material is secreted by the membrane lining the alveoli and its first appearance more or less coincides with the period of viability, i.e. the earliest time at which extrauterine survival is possible. A surface film containing surfactant has the unique property of exerting a high surface tension when it is stretched, but a very low tension when the surface is compressed. As a result the surface tension in a small alveolus is less than that of a larger alveolus and their volumes will tend to equalize, thus preventing lung collapse and promoting alveolar stability. The amount of surfactant in the alveoli is determined by the rate of synthesis in special cells in the alveolar walls and by the rate of release from the cellular stores; cortisol stimulates synthesis and adrenaline stimulates release. Thus the surge of adrenal activity before birth creates a mature, stable lung. Maturation of the lung can be accelerated by stimulating the adrenals with injections of ACTH into the fetus or by administering potent analogues of cortisol to the mother. The latter treatment is highly effective, in infants and young animals about to be born prematurely, in preventing respiratory distress syndrome, a potentially lethal disorder caused by deficiency of surfactant.

Carbohydrate and fat reserves

During the latter part of gestation the fetus of many species accumulates large amounts of glycogen, particularly in the liver. As long as adequate supplies of glucose can reach the fetus from the mother, the total amount of carbohydrate stored in the various tissues appears to depend mainly on the level of fetal adrenal activity. In several species, experimentally induced hypofunction of the adrenals is associated with a big reduction in liver glycogen content which can be restored by administering a corticosteroid. Moreover, when ACTH or corticosteroids are injected into normal fetuses, liver glycogen content becomes abnormally high. The enhanced rate of storage that normally occurs preparatory to birth is probably related in part to accentuated adrenal activity near term.

The pronounced tolerance of newborn animals to severe oxygen deficiency (hypoxia) is attributable, amongst other things, to the high concentration of glycogen present in cardiac muscle. This glycogen store allows the heart to continue its contractile activity in the absence of oxygen, by means of anaerobic glycolysis. The brain is in a less fortunate position, having no glycogen reserves and having to rely on circulating blood glucose. Neonatal hypoglycaemia is a not uncommon condition in human beings and is usually seen in babies who show signs of intrauterine growth retardation. These babies are born with abnormally low levels of liver glycogen, presumably because they have been deprived of glucose by poorly functioning placentas. The brain may be irreparably damaged during episodes of hypoglycaemia, especially if they are accompanied by hypoxia.

Storage of fat is regulated by insulin rather than by corticosteroids. When nutrition is good and blood glucose is maintained at relatively high levels, the action of insulin ensures that glucose available after the requirements of growth are fully met is diverted into fat stores. But when glucose supplies are restricted, the needs of growth have priority over storage and the newborn animal has an emaciated appearance.

As well as the usual white fat, most fetuses have substantial deposits of a peculiar form of adipose tissue called brown fat (Fig. 4.7). Whereas white fat liberates fatty acids and glycerol into the circulation to be metabolized in tissue elsewhere, brown fat is oxidized and releases its energy in the form

Fig. 4.7. Major sites of energy stores in the fetus. Liver glycogen (hatching) provides for the immediate needs of the newborn baby but may be exhausted rapidly. White fat (light stipple) is then metabolized until feeding is established. Brown fat (heavy stipple) acts as a source of heat thus sparing glycogen.

of heat largely within itself. It plays an important part in maintaining body temperature after birth when the newly born animal is no longer within the neutral thermal environment of the mother.

Initiation of lactation

Initiation of lactation

In most mammals, man being an exception, the onset of lactation precedes labour by a variable number of days, thereby enabling the young animal to obtain substantial quantities of food at birth. The mechanisms controlling the onset and maintenance of lactation are fully described in Book 3. But it should be emphasized at this point that the hormonal changes responsible for the onset of lactation emanate from the fetus and placenta and thus can be regarded as a further example of preparation by the fetus for its birth. The fetal mechanisms controlling the initiation both of lactation and of parturition have much in common.

Control of parturition

Fetal genotype

Fetal genotype

The length of pregnancy in a given species is determined by the fetal genotype, the mother generally doing no more than expressing her preference for giving birth at a particular time of the day (see below). When breeds with differing gestation lengths (such as sheep breeds) are crossed so that the fetal genotype contains two differing sets of instructions on pregnancy length, a compromise is found in an intermediate pregnancy length. Even in grey kangaroos in which the duration of pregnancy is little more than the life span of the corpus luteum of a non-pregnant cycle, cross-breeding experiments show that the fetal genotype influences the length of pregnancy. While it is likely that the dominance of the genotype in determining the species-specific duration of pregnancy is universally true in mammals, the manner in which the genotype is expressed varies greatly among species and can be described in any detail in few of them. Only recently has the importance of the conceptus been appreciated and techniques developed to study the fetus in an undisturbed intrauterine environment. The remainder of this chapter will be devoted to the mechanisms underlying the initiation of parturition and to trying to find common threads that bring species together into broad groups.

Prostaglandins

Prostaglandins

The long chain of events leading to parturition in the various species has not only a single origin – the fetal genotype – but it has also a single ending – the release of prostaglandins and its consequences. We will consider first of all how and where prostaglandins are synthesized, then their actions, and finally how the conceptus controls the synthesis and release of prostaglandins.

Prostaglandins are a family of long-chain fatty acids first identified in human semen nearly 40 years ago. Their structure is unusual in that part of the carbon skeleton forms a ring which helps them to dissolve in either lipid or water (Fig. 4.8; see also Book 7 in the First Edition). This property permits prostaglandins to move freely through extracellular water or lipid membranes of cells, and makes them well suited for their special role as local hormones. Prostaglandins have diverse functions in every tissue of the body; in general, they act and are degraded in the same or contiguous tissue in which they are synthesized, any excess escaping into the circulation being totally destroyed during a single passage through the lungs. Luteolysis controlled by prostaglandin $F_{2\alpha}$ ($PGF_{2\alpha}$) released from the uterus in many species (see Book 3) is an apparent exception to this rule. But, in fact, the local nature of the action of $PGF_{2\alpha}$ is preserved by a specialized vascular system that transfers $PGF_{2\alpha}$ directly through the wall of the uterine vein into the ovarian artery and thence to the corpus luteum.

The major prostaglandins formed by the uterus are $PGF_{2\alpha}$ from the endometrium and prostacyclin (PGI_2) from the myometrium. Prostacyclin is a potent vasodilator and probably is concerned with maintaining placental vascular perfusion, whereas $PGF_{2\alpha}$ is the important prostaglandin of parturition. The source of $PGF_{2\alpha}$ during labour is both endometrium and myometrium but the release of $PGF_{2\alpha}$ heralding labour is of endometrial origin.

Actions of prostaglandins. Successful parturition depends not only on contractions of the uterine smooth muscle but equally on changes in the collagen fibres of the inelastic connective tissue of the cervix that cause it to become sufficiently distensible to permit passage of the fetus without damage. Indeed, one could say that the collagen is more important than the smooth muscle since on occasion babies are born without pain or contractions, yet birth cannot occur naturally regardless of the strength and frequency of contractions unless the cervix softens. Normally, softening of the cervix and uterine contractions occur hand-in-hand because both are responses to the local release of prostaglandins. In addition prostaglandins stimulate the smooth muscle cells to develop special areas of contact, gap junctions, which convert the muscle into a network over which electrical impulses readily pass to stimulate co-ordinated contractions. Finally, in species in which pregnancy is maintained to term by progesterone from the corpus luteum, $PGF_{2\alpha}$ from the endometrium contributes to the initiation of luteolysis.

Control of Prostaglandin synthesis. It must be clear by now that to understand what stimulates the release of $PGF_{2\alpha}$ at term is to understand the mechanism initiating labour. Since the rate of synthesis of $PGF_{2\alpha}$ is controlled by the enzyme phospholipase A_2 (Fig. 4.8), which releases arachidonic acid from the abundant stores in phospholipids, the problem can be rephrased: what activates phospholipase A_2? In certain species, this is readily answered – an increase in the oestrogen:progesterone ratio, either from a fall in progesterone or a rise in oestrogen (or commonly both changes). But in other species, parturition begins in the absence of a change in the ratio (Fig. 4.9). To complicate matters further, the source of oestrogen and progesterone near term may be either the placenta or the corpus luteum, and when the placenta is the source the prepartum pattern of progesterone levels may be static, falling or rising. Some sense can be made of this if it is remembered that the only known mechanism for reducing the rate of secretion of progesterone by the placenta is to metabolize it, primarily by hydroxylation at the carbon-17 atom. This leads to a classification of species shown in Table 4.1. The placental enzyme, 17α-hydroxylase, is activated by cortisol in the fetal (but not maternal) circulation (Fig. 4.10). Important implications stem from this classification. Firstly, the presence of placental 17α-hydroxylase provides a mechanism whereby a prepartum surge of fetal cortisol, whether spontaneous or induced experimentally, can initiate parturition regardless of whether the placenta or the corpus luteum is the source of oestrogen. Conversely, the absence of 17α-hydroxylase predicts that the fetal adrenal, or indeed the

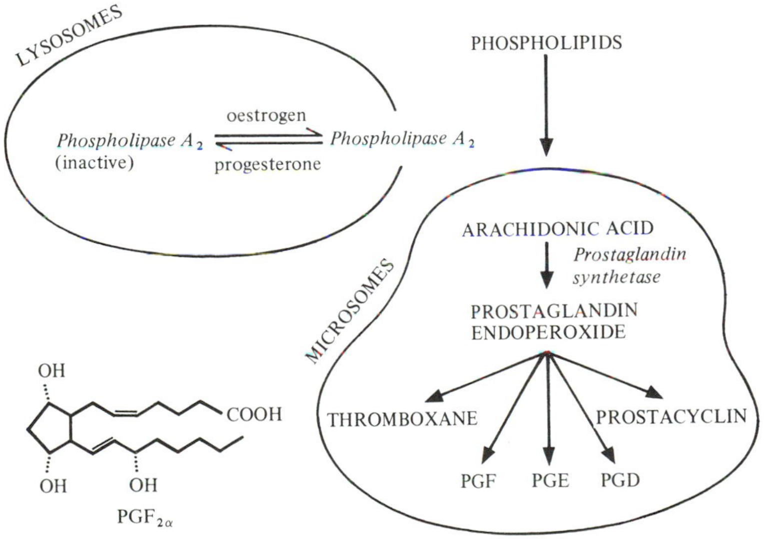

Fig. 4.8. Pathway for synthesis of the main members of the prostaglandin 'family'. Arachidonic acid is liberated from phospholipids by phospholipase A_2 and is converted to prostaglandins by synthetase enzymes that add oxygen to the molecule. The activity of phospholipase A_2 is influenced by oestrogen and progesterone. Anti-inflammatory drugs such as aspirin prolong pregnancy by inhibiting the synthetase enzymes. The structure of $PGF_{2\alpha}$ is shown.

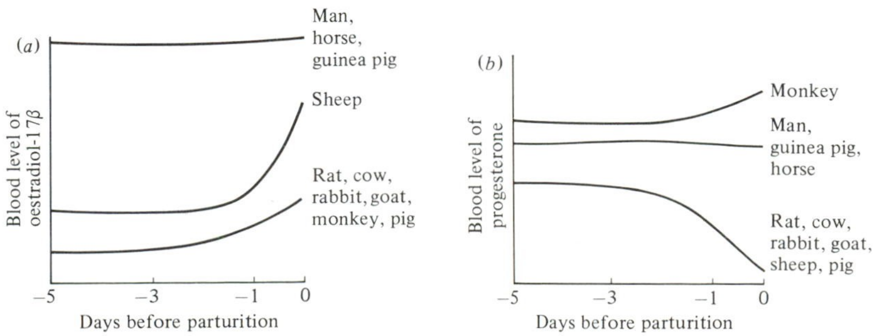

Fig. 4.9. Examples of differing patterns of oestrogen (a) and progesterone (b) in various species showing that the levels in the blood at term may rise, fall or remain constant.

fetus itself, cannot initiate parturition and that the administration of cortisol or ACTH will not stimulate labour.

Secondly, in species in which the placenta is the source of progesterone and placental 17α-hydroxylase is absent, the onset of parturition is not associated with a fall in progesterone levels, nor is the administration of progesterone effective in inhibiting labour. Thirdly, the presence of the enzyme provides the basis for elevated oestrogen levels before parturition, because 17α-hydroxyprogesterone can be readily converted to oestrogen by the placenta of many species (Fig. 4.10). Conversely, in the absence of the enzyme, the placenta can synthesize oestrogen only when there is a 'feto-placental unit' in which the fetus supplies a ready-made precursor from the fetal zone of the adrenal (see Book 3). In such pregnancies, a sharp rise in oestrogen levels before parturition is unlikely since it would necessitate heightened activity in the fetal zone of the adrenal. Finally,

Table 4.1. *Classification of species according to the source of progesterone in late pregnancy and the presence or absence of placental 17α-hydroxylase*

Source of progesterone	Placental 17α-hydroxylase	Examples of species
Corpus luteum	Present	Goat, pig
	Absent	Rat, rabbit
Placenta	Present	Sheep, cow
	Absent	Primates, guinea pig

Fig. 4.10. Diagram showing how fetal cortisol diverts progesterone into the synthesis of oestrogen by stimulating placental 17α-hydroxylation. Thus progesterone levels fall and oestrogen levels rise. This system works in species such as sheep in which the placenta contains 17α-hydroxylase, but not in species such as man in which the enzyme is lacking.

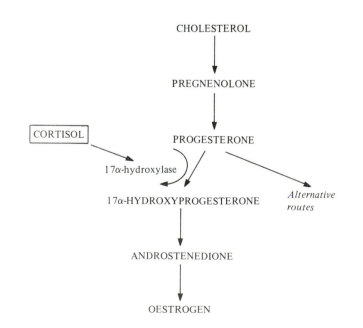

luteolysis in corpus luteum-dependent species lacking placental 17α-hydroxylase may be initiated by withdrawal of a placental luteotrophin rather than by release of uterine $PGF_{2\alpha}$.

Species differences

The mechanism of initiation of parturition in sheep, an example of species in which the placenta is the source of progesterone and contains 17α-hydroxylase, is shown in Fig. 4.11, which also illustrates some of the experimental procedures that helped to develop this model. Figs. 4.10 and 4.11 should be viewed together since the former completes the relationship between the fetal adrenal and placental metabolism. The goat, although a close relative of the sheep, depends on the corpus luteum rather than the placenta as the source of progesterone. Nevertheless, the mechanism is very like that of the sheep with the exception that activation of placental 17α-hydroxylase enhances metabolism of progesterone coming from the corpus luteum rather than the placenta, and the resulting oestrogen-stimulated release of $PGF_{2\alpha}$ causes luteolysis first and uterine activation second.

Because of difficulties in studying the fetal compartment of the small polytocous species such as rats and rabbits which have numerous fetuses and short gestation lengths, much less is known of how their conceptuses

Fig. 4.11. Diagrammatic summary explaining how the fetal lamb controls the onset of labour. Experimental procedures that lengthen or shorten pregnancy are shown. (Adapted from G. C. Liggins. In *Foetal Autonomy*, ed. G. E. W. Wolstenholme and M. O'Connor. Churchill; London (1969).)

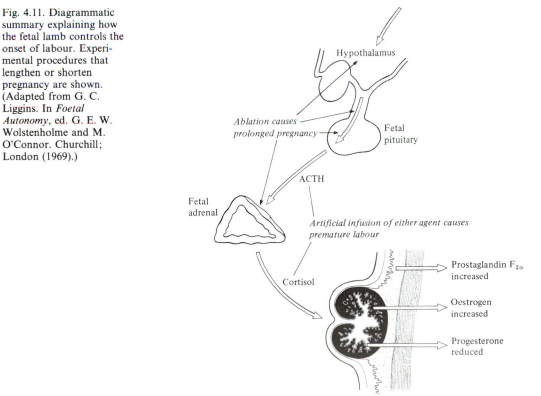

control parturition than in the larger farm animals. Luteolysis precedes parturition but the luteolytic signal is uncertain. Oxytocin, which has only a minor role, limited to the expulsive phase of parturition in many other species, is released early in parturition in rabbits and may be important in activating the uterus both directly and by stimulating the release of $PGF_{2\alpha}$ from the endometrium.

The remaining group of species including man, subhuman primates and guinea pigs have placentas that synthesize progesterone and lack 17α-hydroxylase. Parturition begins without measurable changes in any of the pregnancy hormones except prostaglandins, and no hormonal manipulations other than the administration of prostaglandins or the inhibition of their synthesis affect the length of pregnancy. Strong circumstantial evidence exists to support the conceptus (including fetus, placenta and membranes) as the determinant of pregnancy length, and also the release of $PGF_{2\alpha}$ as the means of effecting parturition. But how the conceptus communicates with the tissues from which $PGF_{2\alpha}$ will be released when the time is ripe remains a mystery. Research is currently directed towards testing the hypothesis that the signal emanates from the chorion (trophoblast and chorionic membrane) and acts directly on the adjacent endometrium where it activates phospholipase A_2. The hypothesis gets some support from the obstetrical manoeuvres used to induce labour in women; the widely differing techniques have in common the physical disruption of the contact between the chorionic membrane and the endometrium close to the cervix.

The role of relaxin, a peptide pregnancy hormone with structural similarities to insulin, remains something of an enigma. It is a product mainly of the corpus luteum and to a lesser degree of the endometrium. Relaxin, as the name suggests, relaxes both smooth muscle and connective tissue, and could be involved in uterine quiescence through pregnancy and cervical softening at term. In some species, notably the pig, large quantities of relaxin are liberated at the onset of luteolysis; in many other species, however, there is little change in levels of relaxin at the onset of parturition.

Circadian rhythms of birth

Not only the day of birth but also the time of birth has a physiological controlling mechanism. For example, eight or nine out of every ten foals are born between 7.00 p.m. and 7.00 a.m. Rats, on the other hand, prefer daylight hours and most are born during 14 hours of light even when the photoperiod is shifted artificially forwards or backwards by 8 hours. A remarkable degree of control is exerted by the alpaca living high in the Andes; nearly all the young are born during the first few hours of daylight, giving them the best chance to get warm and dry before the coldness of night. But on a cloudy, sunless day, birth may be deferred until the following day. Similar, though much less pronounced, circadian rhythms in the time of birth can be recognized in women, pigs, sheep, mice and

Chinese hamsters. The purpose of the rhythms is lost in evolutionary history, but no doubt favoured survival of the newborn animal, perhaps by offering protection from predators. Whether the rhythm arises in the fetus or the mother is uncertain; since the rhythm depends on distinguishing light from darkness it is more likely to be determined by the mother. Neural mechanisms allow many mammals to suppress labour for limited periods of time when circumstances are unfavourable, and the same pathways may mediate the circadian rhythm. One could say that the fetus may determine the day of birth, but the mother may determine its time.

In several ways our ideas about prenatal existence and the meaning of birth have changed radically in recent years. The notion of a dormant fetus springing to life upon entry into the world clearly needs as much revision as Wordsworth's thought that 'Our birth is but a sleep and a forgetting'. Development is remarkably complete before birth, both in form and function, and preparedness is the keynote.

Suggested further reading

Fetal breathing movements, a natural history in animals and man. G. S. Dawes. *Contributions in Gynecology and Obstetrics*, **6**, 62–5 (1979).

Adrenocortical-related maturational events in the fetus. G. C. Liggins. *American Journal of Obstetrics and Gynecology*, **126**, 931–9 (1976).

Timing of the photoperiod and the hour of birth in rats. D. W. Lincoln and D. G. Porter. *Nature, London*, **260**, 780 (1976).

Initiation of parturition. G. C. Liggins. *British Medical Bulletin*, **35**, 145–50 (1979).

Principal substrates of fetal metabolism. F. C. Battaglia and G. Meschia. *Physiological Reviews*, **58**, 499–527 (1978).

Somatomedins and other growth factors in fetal growth. M. W. Brinsmead and G. C. Liggins. *Reviews in Perinatal Medicine*, **5**, 207–242 (1979).

Factors regulating the growth of the placenta: with comments on the relationship between placental weight and fetal weight. G. Alexander. In *Abnormal Fetal Growth: Biological Bases and Consequences*. Ed. F. Naftolin. Abakon Verlagsgesellschaft; Berlin (1978).

The regulation of fetal growth. P. D. Gluckman and G. C. Liggins. In *Fetal Physiology and Medicine*. Ed. R. W. Beard and P. W. Nathanielsz, Marcel Dekker; New York (1981).

Development of the fetal lung. G. C. Liggins and J. A. Kitterman. In *The Fetus and Independent life*. Ed. M. O'Connor. Ciba Foundation Symposium. Pitman Medical Ltd; London (1981).

Chromosomal and other genetic influences on birth weight variation. P. E. Polani. In *Size at Birth*. Ed. K. Elliott and J. Knight. Ciba Foundation Symposium. Associated Scientific Publishers; Amsterdam (1974).

Foetal and Neonatal Physiology. G. S. Dawes. Year Book Medical Publishers; Chicago (1968).

5

Pregnancy losses and birth defects

PATRICIA A. JACOBS

Ideally every successful union of a male and female gamete would result in the birth of a normal healthy offspring who would live to maturity and in due course reproduce and give birth to further normal offspring. In reality, in addition to the enormous loss of gametes prior to fertilization there is a surprisingly high rate of post-fertilization mortality in all mammals in which reproduction has been adequately studied. However, in no species is it as evident or as well studied as in man. Because more is known about reproductive loss in human beings than in any other mammal, most of the information in this chapter is based on our own species. However, there is every reason to suppose that the causes of reproductive loss in other mammals are similar to those in man, although the relative importance of any one mechanism may differ widely from species to species.

Reproductive loss can occur at any time in pregnancy, from the earliest weeks to the time of birth. The fetuses lost during pregnancy are often abnormal and, had they survived, would have had severe birth defects. Therefore, in terms of causation there is no clear distinction between pregnancy losses and birth defects, although the consequences to parents and to society of these two types of adverse pregnancy outcome are very different.

Causes of pregnancy wastage and birth defects
Broadly speaking all pregnancy losses and birth defects are the result either of a genetic abnormality in the fetus or child, or an adverse environment. The latter can be the result of suboptimal conditions for fetal growth and development as a consequence of inadequacy or abnormality of the mother's reproductive system, or the result of exposure of the fetus to an environmental insult, such as an infectious agent or a harmful drug. While both genetic and environmental factors are associated with losses at all stages of pregnancy and with birth defects, their relative importance is different at different stages of pregnancy.

Before dealing with specific causes for early, mid- and late pregnancy losses and birth defects, we must give some thought to the types of genetic and environmental factors associated with abnormal products of pregnancy.

Chromosomal abnormalities

Every organism has a specific number of chromosomes and these chromosomes contain a characteristic array of genes which control all processes of growth and development. Any gain or loss of chromosomal material almost invariably results in gross abnormalities of the fetus which prejudice its survival. Gain or loss of chromosomal material occurs as either a numerical abnormality – that is, the presence or absence of one or more chromosomes – or as an abnormality of size or shape of one or more chromosomes.

Numerical abnormalities occur in the progeny of parents who are themselves quite normal. The abnormality arises as a chance mistake during formation of the egg or spermatozoon, or more rarely during fertilization or at a very early cleavage division of the zygote (Fig. 5.1). Detailed comparison of parental and fetal chromosomes has shown that errors giving rise to chromosomally abnormal gametes usually occur at the first meiotic division in the mother. The reasons for this are not understood

Fig. 5.1. A diagram to show how chromosomes 21 from normal parents can become distributed in their normal offspring and in four possible types of trisomy 21 offspring. The short arm and satellite regions of these chromosomes look different in different people and so can be recognized; in addition, they are stable and inherited, and therefore can be used to trace the origin of chromosomal abnormalities. In the trisomy 21 offspring illustrated, types (*a*) and (*b*) have inherited one chromosome 21 from the father and two from the mother. In (*a*) the two maternally derived chromosomes appear different and their presence together could arise from an error in the first maternal meiotic division, while in (*b*) they appear the same and their presence together is probably due to an error in the second maternal meiotic division. In (*c*) and (*d*) the additional chromosome 21 is paternal in origin. In (*c*) it is here because of an error of the first paternal meiotic division, and in (*d*) because of an error in the second meiotic division.

and the only variable known to be strongly associated with the production of chromosomally abnormal gametes is advancing maternal age. After the age of 35 or so the probability of a woman producing a chromosomally abnormal egg increases dramatically (Fig. 5.2).

Structural abnormalities are the result of chromosomes breaking and rejoining in an abnormal way. Such errors may occur during the first meiotic division when chromosomes are normally breaking and rejoining in the process of crossing-over. Various chemicals and X-rays have been shown to produce structural abnormalities of the chromosomes in the germ cells of laboratory animals. However, to date, exposure to harmful chemicals and X-rays has not been unequivocally associated with the production of structurally abnormal chromosomes in the germ cells of man.

Structural errors in chromosomes are the one type of chromosomal abnormality that may be inherited. Parents who are themselves normal but who carry a balanced structural rearrangement, that is one in which there is neither gain nor loss of genetic material, may have an increased risk of producing a fetus with an unbalanced form of the same arrangement, that is one where there is gain or loss of genetic material (see Chapter 1, Fig. 1.7). About half of all conceptuses having an unbalanced structural abnormality of the chromosomes inherit it from a parent who has a balanced form of the abnormality, while in the remaining half the error occurs during the formation of a gamete in a chromosomally normal parent.

Gametes containing an abnormal amount of chromosomal material give rise to conceptuses with unbalanced chromosome complements which almost always die and are subsequently aborted spontaneously (Table 5.1). Occasionally, chromosomal abnormalities are compatible with survival to birth but the child usually shows multiple birth defects.

Single-gene defects

Reproductive losses and birth defects can result from abnormalities at the level of a single gene; that is, a part of a chromosome governing the production of a single protein. Abnormal genes may be on the X chromosome, in which case they affect mainly males and are carried by

Fig. 5.2. (*a*) The incidence of spontaneous abortion increases with increasing maternal age, especially at maternal ages greater than 35 years. Most of this is due to the increased incidence of chromosome abnormalities in the eggs of older women. ●, all spontaneous abortions; ○, chromosomally abnormal abortion. (*b*) The incidence of Down's syndrome at birth (trisomy of chromosome 21) increases with increasing maternal age, especially at maternal ages greater than 35 years. (From J. L. Hamerton. *Human Cytogenetics.* Academic Press; New York and London (1971).)

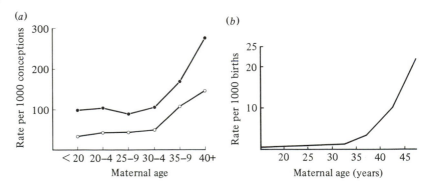

females, or on an autosome, where they affect males and females equally. Autosomal gene disorders may be dominant and affect half the offspring of an affected individual, or they may be recessive and affect one in four children of parents who both carry the abnormal gene. Both parents are more likely to have the same abnormal gene if they are related, and therefore recessive diseases are more common in individuals whose parents are blood relatives. Only a fraction of all single-gene disorders directly affects reproduction, the majority being associated with birth defects rather than pregnancy loss.

Maternal limitations

The most obvious environmental insult to a developing fetus is that provided by an adverse maternal environment. This may result from a structural abnormality of the mother's reproductive tract, such as an abnormally shaped uterus; from a functional abnormality, such as an incompetent cervix; or from a faulty implantation of a blastocyst resulting in a suboptimal position of the placenta. All these types of abnormality are associated with fetal loss or with premature delivery.

Furthermore, maternal nutritional and other habits may have a bad effect on the fetus. Women who smoke or drink even a moderate amount of alcohol during pregnancy are at a somewhat increased risk of having a spontaneous abortion, while a very high proportion of children born to women who are chronic alcoholics are affected with the 'fetal alcohol syndrome', having microcephaly, mental retardation and growth deficiency.

Environmental agents

Few environmental agents are known to be associated with fetal loss in man, although bacterial infections can lead to serious prenatal losses in

Table 5.1. *Percentage incidence of chromosomal abnormalities in 921 human spontaneous abortions with abnormal karyotypes*

Abnormality	Chromosome number	Percentage of total
Monosomy	45	24.2
Trisomy	47	44.5
Double trisomy	48	1.9
Triploidy	69	15.1
Tetraploidy	92	7.1
Mosaicism	—	2.6
Translocations	46	4.3

Source: From Hassold *et al.*, *Ann. Hum. Genet.* (*Lond.*) **44**, 151–178 (1980).

farm animals. However a number of environmental agents, notably viruses, X-rays and drugs, are known to be teratogenic, that is they cause birth defects following exposure *in utero*. The best known viral teratogen is the German measles virus (Rubella) which produces severe birth defects if fetuses are exposed to it during the first 3 months of gestation. We have long known that large doses of X-rays early in pregnancy may result in microcephaly and mental retardation, and there is increasing evidence that many drugs taken during pregnancy are harmful to the developing fetus. The best-documented teratogenic drug is Thalidomide, which caused gross abnormalities of the limbs in children born to women taking the drug during pregnancy.

Some chemicals taken during pregnancy can affect the fetus in very subtle ways and the resulting defects may not be apparent for many years or even decades. Perhaps the most chilling example of this yet known is the long-term effect of diethylstilboestrol which was administered widely to pregnant women in the United States during the 1950s and 60s because it was thought to prevent spontaneous abortions and premature deliveries. Only in the 1970s did people begin to realize that female children born to mothers who took the drug had an increased rate of a rare type of malignancy of the uterine cervix. Now in the 1980s it is becoming apparent that some men and women who were exposed to the drug *in utero* are having reproductive difficulties because of abnormalities of the gonads and reproductive tract. The abnormalities in men include undescended testes and reduction or absence of spermatozoa, and in women abnormalities of the vagina and uterus, and a greater likelihood of an abnormal pregnancy, such as ectopic pregnancy, premature labour, or spontaneous abortion.

Pregnancy loss

A pregnancy can terminate at any time from the zygote to birth. For convenience we shall consider such losses as occuring at three periods: (*a*) very early, that is from conception to the time of clinical recognition at 4 to 6 weeks of gestation, (*b*) during early and mid-gestation, that is from clinical recognition to the end of the second trimester, and (*c*) during late gestation, that is in the third trimester. The causes of late pregnancy loss, with the exception of the hazards of the birth process itself, are also implicated in the aetiology of birth defects.

Very early pregnancy loss

Evidence for pregnancy loss in man prior to the time the pregnancy becomes clinically apparent is of two kinds. Firstly, the examination of ovulated eggs and early zygotes recovered during operation showed that 50 per cent appeared grossly abnormal and unlikely to result in a viable pregnancy. More recently the measurement of human chorionic gonadotrophin levels in blood or urine has made it possible to detect pregnancy as early as 9–12 days after conception and it has been suggested that about 30 per cent of women who react positively to these tests at 10 days of gestation lose their fetus before their pregnancy becomes clinically recognized. When this figure is added to the 25 per cent of women whose pregnancies abort spontaneously subsequent to their being clinically recognizable, we can conclude that about 50 per cent of all human conceptuses that implant successfully are subsequently lost. This figure of course does not take account of pregnancies that are lost prior to implantation. While there is no information on the magnitude of pre-implantation losses in man, we believe these are not insubstantial; the probability of producing a full-term infant following intercourse and ovulation in a normal menstrual cycle is only about 25 per cent. These early losses might result not only from abnormalities of the early conceptus itself but also from hazards faced on its journey to the uterus, and failure of the uterine endometrium to provide a suitable milieu for implantation.

At present we do not know the causes of early post-implantation pregnancy losses but there is every reason to suppose that they are similar to those causing wastage later in pregnancy. It is likely that certain genetic causes, especially loss of whole chromosomes, are responsible for many of the early pregnancy losses.

Early and mid-pregnancy loss
Careful studies show that about 25 per cent of all human pregnancies are lost by spontaneous abortion during the first and second trimesters of a clinically recognized pregnancy. During this period fetal loss rises to a peak at around 11 or 12 weeks of gestation and thereafter declines. The single most common cause of pregnancy loss during the first and second trimesters is the spontaneous abortion of chromosomally abnormal fetuses. No fewer than 50 per cent of all fetuses spontaneously aborted at this time have abnormal chromosome complements and half of these are trisomic, that is they have 47 chromosomes instead of a normal number of 46 (see Table 5.1). The additional chromosome usually results from an error, called non-disjunction, occurring at the first meiotic division during formation of the egg. The resulting egg with 24 chromosomes instead of 23, after fertilization with a spermatozoon having 23 chromosomes, gives rise to an embryo with 47 chromosomes. Such embryos are almost always grossly abnormal and may be aborted before giving rise to a clinically recognizable pregnancy. While non-disjunction usually occurs during oogenesis (see Book 1, Chapter 2) it can also happen during spermatogenesis (Book 1, Chapter 4), and it can affect any of the 23 chromosome pairs. However it seems to involve some chromosomes more than others. Chromosome 16 in particular, for reasons that are not understood, appears to undergo non-disjunction much more often than the other chromosomes and as a result as many as 1 per cent of all clinically recognized human conceptuses have an additional chromosome 16 (Fig.

5.3). Trisomy for chromosome 16 is evidently highly lethal for it is never found in liveborn children. Chromosome 21 also undergoes non-disjunction fairly frequently, although at less than one-third the frequency of chromosome 16. However while about four out of every five fetuses with an additional chromosome 21 abort spontaneously, about one in five are born alive and they have a constellation of birth defects known as Down's syndrome (Fig. 5.4).

Non-disjunction or a similar error of meiosis in one or the other parent can also give rise to an egg or spermatozoon with one too few chromosomes, resulting in an embryo with 45 instead of 46 chromosomes. Such an embryo is called monosomic because it has one rather than two members of one of the chromosome pairs. With the exception of a fetus that is missing a single sex chromosome, that is with a 45, X complement, monosomic fetuses are virtually never seen among clinically recognized abortions. Lack of genetic material is much more deleterious than additional material, and we assume that though monosomic fetuses occur with at least

Fig. 5.3. A karyotype prepared from a fetal cell that had 47 chromosomes, including an additional chromosome 16. The parental chromosomes 16 are shown inset above, and both they and the fetal chromosome 16 are enlarged to show the variable region which demonstrates that two of the three fetal chromosomes were maternal in origin; as they are different, we can also infer that the error occurred at the first maternal meiotic division.

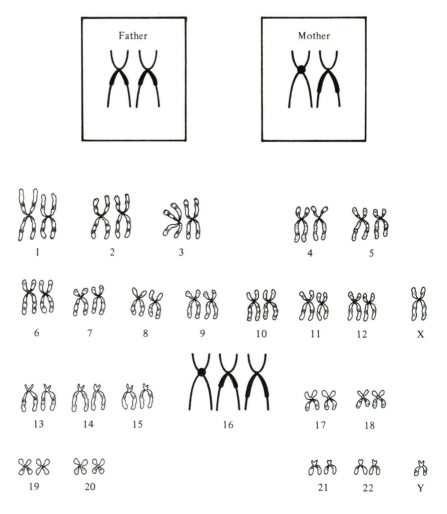

the same frequency as trisomic fetuses, they are so abnormal that they are lost very early in pregnancy. Monosomy is probably one of the main causes of very early pregnancy loss. The only monosomy capable of surviving until pregnancy becomes clinically recognizable is that for the X chromosome, and in fact 45, X fetuses are surprisingly common (Fig. 5.5). They are found in about 10 per cent of all spontaneous abortions, thus representing some 1 to 2 per cent of all clinically recognizable conceptions. About 1 in 5000 liveborn females is found to have a 45, X constitution, such individuals having short stature, 'streak' gonads, and

Fig. 5.4. A 4-month-old baby girl with Down's syndrome (mongolism), caused by the presence of an additional chromosome 21. Distinctive features of the condition include a flat face with widely spaced slanting and rather narrow eyes, and a short skull and neck. (Photographs by courtesy of Professor P. E. Polani.)

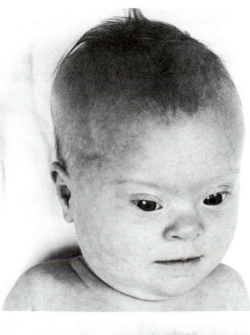

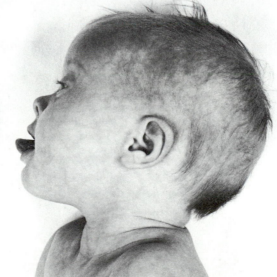

a failure of development of secondary sexual characteristics (see Chapter 3). However their intellect is unimpaired and, in view of the relative normality of the few 45, X conceptuses that do survive to term, it is a mystery why the vast majority should die during the first or second trimester of pregnancy.

The other major category of chromosomally abnormal conceptuses has one or two additional haploid complements of 23 chromosomes. If there is one additional set, the embryo is called a triploid, having 69 chromosomes (Fig. 5.6), and if there are two additional complements the embryo has 92 chromosomes and is called a tetraploid. Together these classes comprise 12 per cent of all clinically recognized spontaneous abortions. Triploids can arise in a variety of ways, and in man about 60 per cent of triploids are a result of dispermy, that is fertilization of an egg by two spermatozoa instead of one (see also Book 1, Chapter 6). Another 10 per cent of triploids result from fertilization with a giant spermatozoon containing 46 chromosomes, which arises from the non-occurrence of one

Fig. 5.5. A 45, X human fetus that was spontaneously aborted when of 11 cm crown–rump length. Note the generalized oedema, and the massive hygromas on each side of the neck. The ovaries are quite normal in appearance at this stage, but would subsequently become vestigial 'streaks' as all the oocytes perish, presumably because of lack of the second X chromosome. (From J. Boué, E. Philippe, A. Giroud and A. Boué. *Teratology*, **14**, 3–19 (1976).)

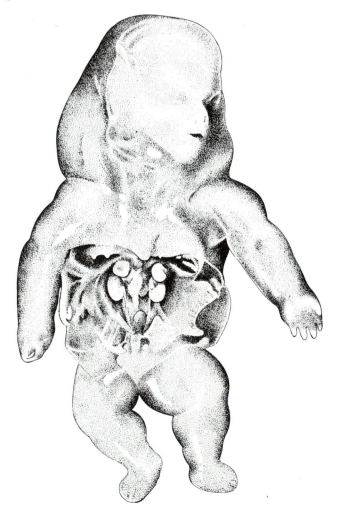

of the paternal meiotic divisions. The remaining 30 per cent of triploids derive from lack of a maternal meiotic division, the resulting egg with 46 chromosomes instead of 23 being fertilized by a normal spermatozoon. Thus triploids can result from an error during the production of a gamete or during the process of fertilization itself (Fig. 5.7). In contrast, tetraploids are formed as a result of an error of cell division in a chromosomally normal zygote: the chromosomes replicate but the cell fails to divide, so

Fig. 5.6. Karyotype of a fetus with 69 chromosomes and an XXX sex chromosome constitution. Such triploids are common in man and represent about 8 per cent of all spontaneous abortions.

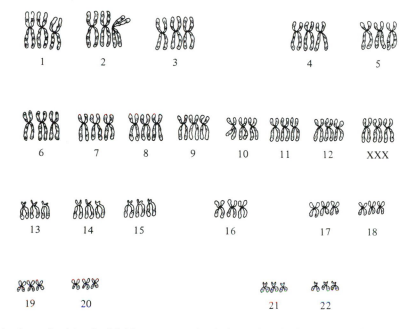

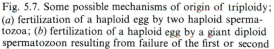

Fig. 5.7. Some possible mechanisms of origin of triploidy; (*a*) fertilization of a haploid egg by two haploid spermatozoa; (*b*) fertilization of a haploid egg by a giant diploid spermatozoon resulting from failure of the first or second paternal meiotic division; (*c*) fertilization of an egg with a giant diploid pronucleus, resulting from failure of the first or second maternal meiotic division, by a normal spermatozoon.

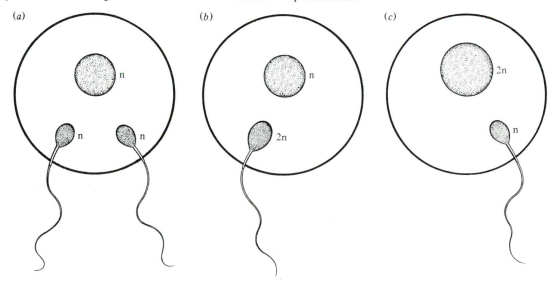

that instead of two cells with 46 chromosomes each, there is one cell that is a tetraploid with 92 chromosomes. Very rarely triploidy is compatible with birth of a liveborn child, but such newborn children have a large number of birth defects and rarely survive more than a few hours. Tetraploidy on the other hand is always lethal at a relatively early stage of gestation and tetraploid abortuses are comprised of a gestational sac in which there is no evidence of fetal development.

Another type of error occurring at fertilization results in a placental abnormality known as a hydatidiform mole. The fetus itself is absent, the placenta is often much enlarged and the villi swollen, the outer layer of

Fig. 5.8. (*a*) Part of a hydatidiform mole, showing the distinctive grape-like clusters of swollen chorionic villi. (From K. Benirschke and S. G. Driscoll. *The Pathology of the Human Placenta.* Springer-Verlag; Berlin and New York (1967).) (*b*) Chromosomes 3, 4 and 13 from a hyda-tidiform mole and its parents. As can be seen, the chromosomes of the mole appear identical with respect to the variable regions. No maternal member of pairs 4 and 13 is present in the mole and in pair 13 one parental chromosome is duplicated.

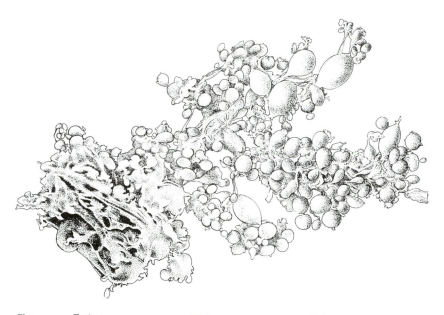

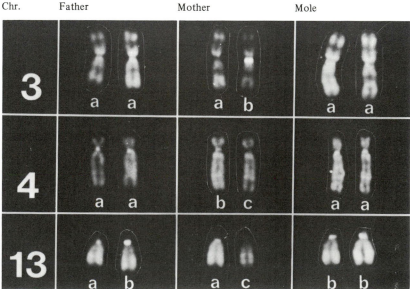

Chr.	Father	Mother	Mole
3	a a	a b	a a
4	a a	b c	a a
13	a b	a c	b b

trophoblast cells being very hyperplastic and secreting large amounts of chorionic gonadotrophin into the maternal circulation – a useful diagnostic feature (Fig. 5.8). Hydatidiform moles have been of great interest and considerable concern for many decades. The interest had sprung in part from variations in the frequency of moles in different localities. In some parts of the world, notably South-east Asia, Japan, the Philippines and Mexico, moles occur in one in 250 clinically recognizable pregnancies, whereas in Europe and North America it is a relatively rare condition, occurring in about one in 2000 pregnancies. The reasons for these different frequencies are not understood. It is known that moles occur much more frequently in women at the extremes of reproductive age, with a particularly high incidence at the approach of the menopause (see Table 5.2), and it has also been suggested that they are more common among women in the lower socio-economic classes. However, these facts alone cannot account for all the difference in frequencies among different populations.

The concern about moles is because they have a particular propensity to become malignant, turning into tumours called chörio-carcinomata which will metastasize to other sites in the body. Until chemotherapy became available, such tumours almost invariably led to the death of the mother, but using cytotoxic drugs it is now possible to achieve a complete cure in almost 100 per cent of cases.

Until recently the origin of these abnormal pregnancies was obscure. However, Todashi Kajii and his colleagues, and ourselves, have recently shown that the great majority of complete hydatidiform moles have a 46, XX chromosome constitution and that all the chromosomes appear to come from the father, the mother making no genetic contribution to the conceptus (Fig. 5.8). The great majority of moles arise as the result of a single normal haploid spermatozoon fertilizing an egg in which the female pronucleus is either absent or non-functional. After fertilizing the abnormal

Table 5.2. *Incidence of molar pregnancy by maternal age*

Maternal age	Approximate incidence of moles per 10000 births
15–19	100
20–24	20
25–29	20
30–34	20
35–39	20
40–44	100
45–49	250
50–54	500

egg, the chromosomes of the spermatozoon double up, thereby restoring the diploid number of 46 and incidentally making the mole homozygous for all paternal genes. If the spermatozoon is Y-bearing, the resulting embryo would have a 46, YY constitution and would die because cells apparently cannot live in the absence of an X chromosome. Occasionally, complete hydatidiform moles may result from the fertilization of an anucleate egg either by a giant diploid spermatozoon, or by two separate haploid spermatozoa. In such cases, the embryo's genome is still entirely paternal in origin but it contains two separate sets of paternal chromosomes (Fig. 5.9). Such moles can therefore have either a 46, XY or a 46, XX constitution. There is some preliminary evidence that moles that are not entirely homozygous are more likely to give rise to chorio-carcinomata than the more common 46, XX homozygous moles.

Complete hydatidiform moles appear to be unique to man, but the reasons for their bizarre origin and pathology are as yet quite unknown.

Another type of chromosome abnormality associated with fetal loss is an unbalanced structural rearrangement. While this type of abnormality

Fig. 5.9. Diagram to illustrate the origin of hydatidiform moles. (*a*) A single haploid spermatozoon fertilizes an oocyte with a defective pronucleus which takes no part in the fertilization process. The chromosomes of the spermatozoon double up, to give a diploid zygote that is therefore homozygous for all paternal genes. The fact that no YY moles are found suggests that at least one X chromosome is essential for life, even in a molar trophoblast. (*b*) Two haploid spermatozoa, either X- and Y-bearing or both X-bearing, fertilize an oocyte with a defective nucleus which takes no part in the fertilization process. The two male pronuclei subsequently fuse to give a heterozygous diploid zygote. This mode of origin accounts for some if not all the XY moles (4 per cent of all moles), and probably also for a small proportion of the XX moles.

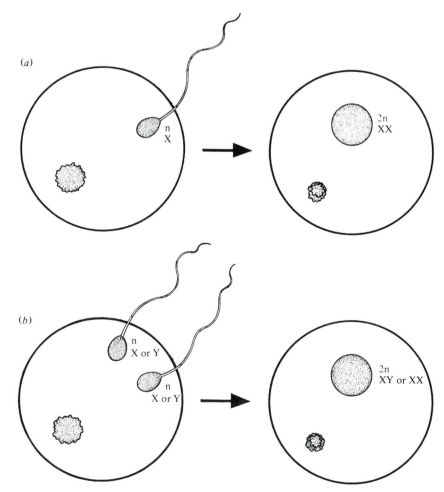

accounts for only about 2 per cent of all spontaneous abortions, it is of considerable importance, because as mentioned earlier, in about 50 per cent of abortions with an unbalanced chromosome constitution one or other parent is found to have a balanced form of the rearrangement. Parents with a balanced structural chromosome abnormality are themselves quite normal in appearance, because they have the correct amount of chromosome material, but they are likely to produce a gamete with an abnormal amount of genetic material, which in turn results in a conceptus with a genetically unbalanced structural rearrangement. While these conceptuses usually abort, they occasionally survive as liveborn children with multiple birth defects. However, in spite of the increased risk of spontaneous abortions or children with birth defects, the majority of offspring of parents with balanced structural rearrangements are quite normal. Furthermore it is now possible, by examining the chromosomes of fetal cells obtained at amniocentesis, to recognize any chromosomally unbalanced fetus at an early stage of pregnancy and to offer the parents the possibility of a therapeutic abortion.

While chromosome abnormalities are the main cause of first and second trimester pregnancy loss, they account for only one-half of such losses, and we still know little about the causes for the spontaneous abortion of chromosomally normal fetuses. Possibly as many as 10 to 20 per cent may be the result of ABO blood group incompatibility between mother and fetus, and some may be the result of abnormalities of the maternal reproductive tract.

The effects of a mother smoking and drinking, even of moderate amounts of alcohol, are beginning to become apparent by the second trimester, and we can reasonably assume that other adverse environmental factors, as yet unrecognized, may also be associated with the spontaneous abortion of chromosomally normal conceptuses.

Surprisingly enough, there is no evidence that single-gene defects account for any substantial fraction of pregnancy loss. There is, for example, no increase in spontaneous abortion associated with any form of inbreeding, thus suggesting that recessive genes are not important in this regard.

Late pregnancy loss

Losses during the third trimester of gestation are relatively rare and occur in only about 1 per cent of pregnancies. Many of the causes of late pregnancy loss are the same as those responsible for the death of children with birth defects in the neonatal period. About 6 per cent of late pregnancy and neonatal losses are due to chromosomal abnormalities. This is much less than the proportion of chromosomal abnormalities found in mid-pregnancy losses, but still represents ten times as many chromosomal abnormalities as are found in random populations of liveborn children. The types of chromosomal abnormalities associated with late pregnancy and neonatal loss are limited to three kinds of trisomy, namely trisomy for chromosomes 13, 18 or 21, and the very occasional triploid conceptus or a 45, X monosomy. While individuals with these abnormalities can survive until birth, they rarely do so, and the great majority die and are spontaneously aborted at about the twelfth week of gestation.

It is among late pregnancy losses that environmental factors take their greatest toll. Abnormalities of structure and/or function of the maternal reproductive tract are usually associated with late pregnancy loss, as is smoking and the consumption of alcohol during pregnancy. Furthermore, harmful environmental agents such as viruses, and many drugs that are known to cause congenital abnormalities in newborn children could also be associated with fetal wastage during the later stages of pregnancy, but so far we have no reliable information on this point.

Birth defects

The later in gestation that a loss occurs, the greater is the impact on the parents and on society. The birth of a defective child is the most disturbing of all adverse pregnancy outcomes, and it is also the one about which we have most information. Birth defects are due to both genetic and environmental causes, and the genetic causes include both chromosomal abnormalities and single-gene defects. While this latter category is an important cause of birth defects, it does not seem to be associated to any great extent with prenatal loss. This restriction of the effect of single-gene abnormalities to the postnatal period may be more apparent than real. It is considerably easier to recognize and characterize single-gene defects in living individuals than in the macerated tissue that is often all that is available from a spontaneous fetal loss. Thus some unexplained fetal losses may well be due to unrecognized single-gene defects.

About 2 per cent of all liveborn babies have a major congenital abnormality that is obvious at birth or shortly thereafter, and about 10 per cent of the abnormal babies have an aberrant chromosome constitution. The majority have an extra chromosome 21, while a few have an extra chromosome 13 or 18. About 20 per cent of all congenitally abnormal newborn children with a chromosomal abnormality have an unbalanced structural chromosome rearrangement, which is inherited in about half of the cases, almost always from the mother. Abnormal numbers of sex chromosomes are found in 1 in 500 newborn children; those affected usually appear normal at birth, although the adverse affects of the abnormal sex-chromosome constitution may become manifest later in life as mental retardation and subfertility.

Some 1500 individually rare disorders have been described that are due to single-gene defects, and about 1 per cent of the population are affected with such abnormalities, the majority of which are recognizable at or around the time of birth. These defects include: haemophilia and some types of muscular dystrophy that are caused by genes on the X chromosome and affect only males; achondroplasia, attributable to a dominant auto-somal gene defect; and cystic fibrosis and Tay–Sachs disease, due to recessive gene defects.

It is becoming obvious that a large, but as yet unknown, proportion of birth defects are the result of environmental insults to the fetus. Perhaps the best known of these is the Rubella syndrome, consisting of deafness, cataracts and heart defects, the result of exposure of the fetus during early pregnancy to the virus responsible for German measles. Thus an infectious agent that causes only a mild disease in the mother may be associated with a very severe abnormality in the fetus. The only other infectious agent that has definitely been shown to be associated with birth defects in man is the cytomegalovirus, which exists in about half of the adult population without producing any ill effect.

Many drugs, including thalidomide, 6-mercaptopurine and dilantin, have been shown to be associated with severe birth defects in children exposed *in utero*. Each drug is associated with a characteristic syndrome, the severity of which depends on the stage of gestation at which the fetus is exposed. While the drugs have remarkably little effect on the mother, they can have a disastrous consequence for the developing fetus. The extreme sensitivity of the fetus to environmental agents that have little effect in adults has only recently been appreciated and almost certainly many other birth defects will be found to be the result of such environmental agents.

The magnitude of human pregnancy loss is surprising, and considerably greater than in other mammals. The reasons for this are not understood but must be complex and include the fact that man is a non-seasonal and non-cyclic breeder. This probably results in the frequent union of suboptimal gametes which may be responsible for much of the early reproductive loss in our species. Furthermore, man continues to breed into relatively old age and this practice alone is responsible for a significant proportion of chromosomally abnormal eggs, which in turn accounts for much, if not all, of the well known increase in spontaneous abortion with increasing maternal age.

Virtually no data exist that allow a comparison of the incidence of birth defects in man with that in other outbred mammalian species. The incidence of birth defects may appear higher in man because only in our own species do any significant proportion of defective progeny survive to become noticed and studied. Knowledge of the causes of reproductive loss and birth defects has accumulated rapidly in the past two decades and with this knowledge comes the possibility of rational prevention, selective therapeutic abortion of abnormal fetuses early in gestation and improved methods of treatment of children born with birth defects.

Suggested further reading

Dispermic origin of XY hydatidiform moles. K. Ohama, T. Kajii, E. Okamoto, Y. Fukada, K. Imaizumi, M. Tsukahara, K. Kobayashi and K. Hagiwara. *Nature, London*, **292**, 551–552 (1981).

Mechanism of origin of complete hydatidiform moles. P. A. Jacobs, C. M. Wilson, J. A. Sprenckle, N. B. Rosenheim and B. R. Migeon. *Nature, London*, **286**, 714–716 (1980).

Recognisable Patterns of Human Malformation. D. W. Smith. In *Major Problems in Clinical Pediatrics*, Vol. 7, 2nd edition. Saunders; Philadelphia (1976).

Physiology and Genetics of Reproduction. Ed. E. M. Coutinho and F. Fuchs. Plenum; New York (1974).

Genetics in Medicine. J. S. Thompson and M. W. Thompson. Saunders; Philadelphia (1980).

Aging Gametes. Ed. R. J. Blandau. Karger; Basel (1975).

Human Embryonic and Fetal Death. Ed. I. H. Porter and E. B. Hook. Academic Press; New York and London (1980).

6

Manipulation of development

R. L. GARDNER

Until quite recently there had been few attempts to manipulate mammalian development, because it is so much more difficult than experimenting with developing frogs and sea urchins. Satisfactory methods had first to be devised for the culture of pre-implantation embryos and their transfer to recipient animals, so-called 'uterine foster-mothers'. Many of the experimental approaches have been adapted from the basic repertoire of embryologists interested in non-mammalian forms since the late nineteenth century. Their application to the mammalian embryo has proved an inviting challenge and endeavours are now beginning to reap very considerable rewards. The aim of this chapter is to explain how some of these procedures have both enhanced our knowledge of early mammalian development and conferred a measure of control over it.

Experimental chimaeras

Perhaps the most dramatic manipulations of mammalian development are those by which usually two, or possibly more, populations of cells carrying prescribed genetic differences are made to coexist in an individual from a very early stage. These products are termed chimaeras after a legendary

Fig. 6.1. The original chimaera, daughter of Echidne, was a composite organism boasting characteristics of the lion, goat and serpent. This is because she symbolized the Tripartite Sacred Year, with the lion representing spring, the goat summer, and the serpent winter. Such a mythological monster clearly has little in common with the normal integrated organisms that feature so prominently in current mammalian embryological research. (From A. McLaren (1976) – see Suggested Further Reading.)

Fig. 6.2. Technique for transferring an inner cell mass to a blastocyst or trophectodermal vesicle. The blastocyst is held by suction on the tip of one smooth glass pipette (*left*) and the donor inner cell mass on another (*right*). Three sharp glass needles are used to make a triangular hole in the trophectoderm wall and overlying zona pellucida, clear of the inner cell mass of the host blastocyst. The donor inner cell mass is then introduced into the blastocoele through this hole. Essentially the same procedure is used for transferring individual cells, except that the right-hand pipette is wider at the tip to accommodate the donor cells within it.

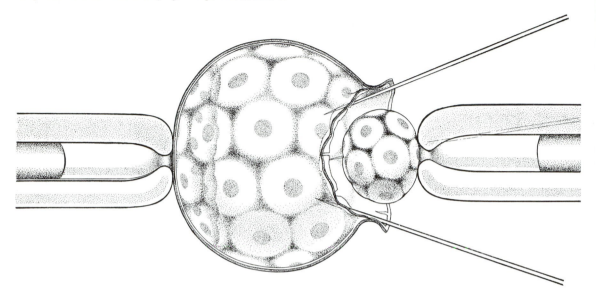

Fig. 6.3. Three successive stages in early development showing the principal tissues of the embryo. (*a*) Free mouse blastocyst at 3½ days *post-coitum* prior to loss of the zona pellucida; (*b*) implanting blastocyst at 4½ days *post-coitum*; (*c*) implanted egg-cylinder stage embryo at approximately 6 days *post-coitum*.

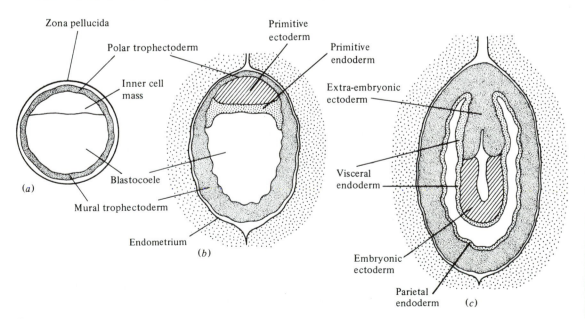

monster (Fig. 6.1). Since experimental chimaeras have been used so extensively in investigations discussed in this chapter, it is appropriate to consider them in some detail at this stage. They may be produced by aggregating pairs of cleaving embryos (as described in Chapter 1) or, as illustrated in Fig. 6.2 by injecting extra tissue or cells into blastocysts. Genetic differences between embryos used to produce such chimaeras provide indelible markers for assessing their cellular contributions to the resulting conceptuses or offspring. The mouse has proved the ideal species for these experimental manipulations because of its short gestation interval and wealth of known genes, although similar chimaeras have also been produced in the rat, rabbit and sheep.

Chimaeras may arise spontaneously in mammals in several different ways, most frequently by exchange of cells between fetuses (as between dizygotic twins in cattle – discussed in Chapter 3) or by passage of cells in either direction across the placenta. Chimaerism is limited in these instances to one or a few tissues, notably blood and trophoblast. A similar degree of chimaerism may be produced experimentally by injecting haemopoietic tissue into neonatal animals whose immunological system is not yet mature, or by applying like measures to adults treated with X-rays or immunosuppressive drugs.

It is important to appreciate that the situation is very different in chimaeras produced experimentally at or before the blastocyst stage, and in those occasional natural chimaeras that arise either by spontaneous embryo aggregation, or by double fertilization after 'immediate cleavage'. Here all or most tissues, including the germ cell line, may contain a mixture of cells, apparently because the cells come together before they have become committed. Nevertheless, not all chimaeric blastocysts give rise to offspring displaying both cell lines: in some cases one or other line contributes only to extraembryonic tissues or is lost altogether.

Since the mixing of cells precedes differentiation of the immune system, experimental chimaeras may show 'intrinsic' tolerance. This means that provided the constituent genotypes are derived from different inbred strains of mice, grafts from individuals of either parental strain will be retained indefinitely. Grafts from an unrelated strain are rejected as vigorously as by any conventional mouse. The most astonishing feature of these chimaeras is, however, that they develop into normal adults which can provide unique information about their developmental history. The number and diversity of genetic differences that can be united in a single embryo is under the control of the investigator, and is in principle almost unlimited.

Manipulation of the mouse blastocyst

Experimental isolation of blastomeres and the aggregation of cleaving embryos have already been discussed in relation to analysis of the initial differentiation in the early embryo (see Chapter 1). The mechanism of differentiation of trophectoderm versus inner cell mass remains obscure, but the cells of the embryo are clearly partitioned into these two tissues when it begins to form a blastocoelic cavity. Within the next 24 hours of development both the trophectoderm and inner cell mass have undergone subdivision into two populations of cells (Fig. 6.3). How do the properties of these tissues differ, and what is their normal fate later in development? Does the inner cell mass play any part in implantation? How do the two primary tissues of the blastocyst interact in subsequent development? One must investigate the blastocyst by microsurgery in order to answer these and related questions.

Comparison of isolated trophectoderm and inner cell mass tissue

The $3\frac{1}{2}$-day mouse blastocyst is roughly 0.1 mm in diameter and has therefore to be dissected under high magnification with the aid of micromanipulators. Blastocysts can be immobilized by suction on the flame-polished tip of a micropipette and then cut in two, close to the inner cell mass, with a microscalpel. This provides a source of mural trophectoderm free of any enclosed cells of the inner cell mass (Fig. 6.4). Inner cell mass tissue can be obtained by tearing blastocysts open with fine pointed glass needles and stretching the trophectoderm out as a sheet. The inner cell mass is then scraped gently off the trophectoderm (Fig. 6.4). The inner cell mass of the late blastocyst can be further separated into primitive endoderm versus primitive ectoderm in a similar way (see Fig. 6.3 for details of the location of various tissues in early mouse embryos).

More recently, Davor Solter and Barbara Knowles in Philadelphia have developed an elegant and simple 'immunosurgical' technique for isolating inner cell masses from blastocysts. It entails incubation of intact blastocysts in a rabbit antiserum raised against mouse tissue. The antibody molecules are prevented from entering the blastocyst by the specialized junctions between trophectoderm cells, and can therefore bind only to the outer surfaces of the latter. Thus, providing treated embryos are rinsed to remove excess unbound antibody, selective lysis of the trophectoderm cells can be obtained by exposing them subsequently to complement.

Isolated trophectoderm and inner cell mass tissue differ strikingly in properties, both in culture and following transfer to the uteri of pseudopregnant mice. Some major differences between the tissues are summarized in Table 6.1. As indicated, secretion of blastocoelic fluid and induction of implantation are properties peculiar to the trophectoderm, independent of inner cell mass tissue. However, as we shall see later, the inner cell mass appears to play a decisive role in controlling the growth of the trophectoderm.

Vesicles or fragments of mural trophectoderm implant in recipient uteri with a similar frequency to intact blastocysts. They evoke typical decidual swellings in the endometrium, but these lack embryo, amnion, allantois and yolk sac, a finding that accords with other evidence of an inner cell mass origin for these structures (see later). Unexpectedly, such implantation sites also show a very conspicuous lack of trophectoderm derivatives and contain, at most, approximately two-dozen trophoblastic giant cells. Failure to proliferate cannot be attributed simply to a deficiency of cells in the transferred fragments, since partial blastocysts can develop into fetuses or young (Fig. 6.5), though they may contain fewer trophectoderm cells. Two differences distinguish mural trophectoderm fragments that fail to proliferate and intact or partial blastocysts which exhibit normal growth of trophoblast: the former lack both inner cell mass and overlying polar trophectoderm cells (see Figs. 6.3 and 6.4). In order to determine which of these groups of cells is needed for sustaining trophoblast growth, we must examine the development of blastocysts that have been reconstituted from genetically dissimilar mural trophectoderm and inner cell mass tissue.

Development of 'reconstituted' blastocysts

The procedure for reconstituting blastocysts is illustrated schematically in Fig. 6.6. Isolated fragments of mural trophectoderm are cultured until they

Fig. 6.4. Operation to separate trophectoderm and inner cell mass. (*a*) Blastocysts are cut in two along line A–A to obtain pure trophectoderm. (*b*) Blastocysts are dissected open to make possible isolation of enclosed inner cell mass tissue (stipple).

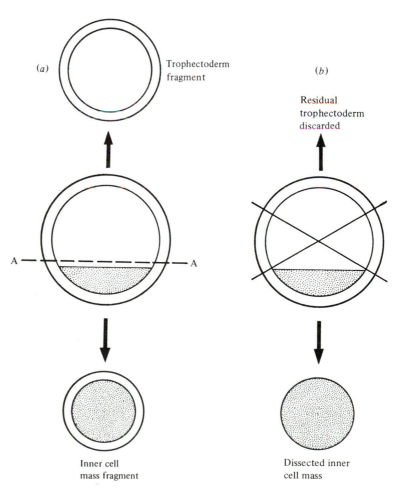

Table 6.1. *Summary of differences between isolated inner cell mass and trophectoderm*

Type of tissue	Secretion of blastocoelic fluid	Aggregation with fragments of like tissue	Initiation of implantation in recipient uteri	Phagocytosis of small particles
Trophectoderm	+	−	+	+
Inner cell mass	−	+	−	−

Table 6.2. *Maximal distribution of clones obtained following injection of $3\frac{1}{2}$-day versus $4\frac{1}{2}$-day inner cell mass (ICM) cells into blastocysts*

	ICM derivatives of host conceptuses									
	Extra-embryonic endoderm		Extra-embryonic mesoderm			Amniotic ectoderm	Fetal			
Type of donor cell clone	Parietal	Visceral	Visceral	Allantoic	Amniotic		Ectoderm	Mesoderm	Endoderm	Germ cells
$3\frac{1}{2}$-day ICM	+	+	+	+	+	+	+	+	+	+
Primitive endoderm from $4\frac{1}{2}$-day ICM	+	+								
Primitive ectoderm from $4\frac{1}{2}$-day ICM			+	+	+	+	+	+	+	+

+, The host tissues that can be colonized by the progeny of a *single* donor cell. Note the reciprocal patterns of chimaerism obtained following transplantation of primitive endoderm versus primitive ectoderm cells. The trophectoderm derivatives are invariably composed of host cells only, in all three classes of chimaeras shown in the Table.

form vesicles, into each of which an inner cell mass is inserted microsurgically by the technique that is used for injecting tissue into blastocysts (Fig. 6.2).

More than half such 'reconstituted' blastocysts that implant yield fetuses, around which trophoblast growth and differentiation has proceeded normally, despite the fact that they lacked polar trophectoderm (Fig. 6.6). Furthermore, the trophoblast expresses the genetic marker characteristic of the trophectoderm donor, and the fetus, amnion, allantois and yolk sac express the genetic marker characteristic of the inner cell mass donor blastocyst. Several conclusions can be drawn from these results. Firstly, isolation and culture of inner cell mass and trophectoderm does not impair their viability. Secondly, an inner cell mass can develop into a normal fetus when surrounded by a genetically dissimilar envelope of trophectoderm. Thirdly, the inner cell mass gives rise to several extraembryonic tissues in addition to the fetus itself. Finally, since the inner cell mass does not make a cellular contribution to the trophoblast, it must clearly be acting indirectly by sustaining division of some mural trophectoderm cells. In

Fig. 6.5. Mouse fetuses and young may be obtained from half blastocysts (*left*) or from blastocyst fragments containing the inner cell mass (*right*).

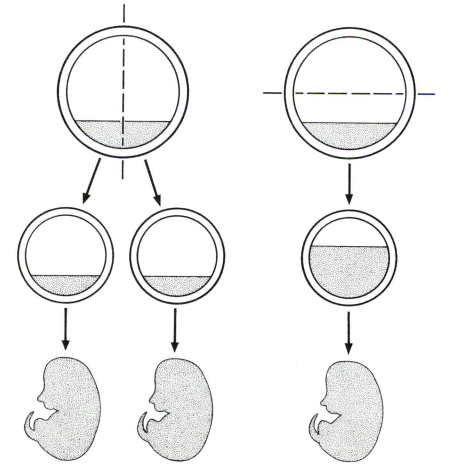

intact blastocysts, all trophectoderm cells except those immediately over-lying the inner cell mass have begun to transform into postmitotic giant cells by $4\frac{1}{2}$ days of development (Fig. 6.3). This argues that the effect of the inner cell mass must be a strictly local one. As discussed earlier, in the absence of inner cell mass tissue, trophectoderm only gives rise to a few giant cells. Together, these various findings suggest that all trophectoderm cells are destined to transform into terminal giant cells soon after they differentiate in the blastocyst, unless they remain adjacent to the inner cell mass. The role of the latter thus seems to be to promote division or prevent transformation of overlying (polar) trophectoderm, and thereby maintain a pool of proliferating cells from which all later trophoblasts take origin.

Andrew Copp, working in Oxford, has obtained clear evidence that mural trophectoderm grows by recruitment of polar cells in the late blastocyst. He has suggested, furthermore, that this recruitment ceases once the mural trophectoderm is firmly attached to the uterine wall, resulting in the local accumulation of polar cells which displace the inner cell mass deep into the blastocoele (see Fig. 6.3). While working in Oxford, Janet Rossant and Leah Ofer found that mitotically active trophoblast tissue of post-implantation embryos also responded to isolation from the

Fig. 6.6. Scheme of experimental 'reconstitution' of mouse blastocyst from genetically dissimilar mural trophectoderm and inner cell mass.

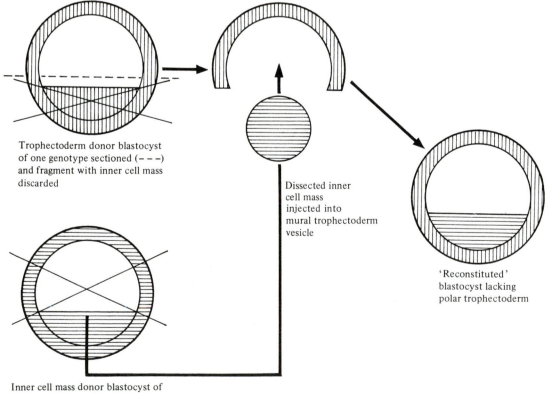

Trophectoderm donor blastocyst of one genotype sectioned (– – –) and fragment with inner cell mass discarded

Dissected inner cell mass injected into mural trophectoderm vesicle

'Reconstituted' blastocyst lacking polar trophectoderm

Inner cell mass donor blastocyst of different genotype torn open and inner cell mass removed – overlying trophectoderm discarded

embryo by prompt cessation of cell division and initiation of giant cell formation. Hence the dependence of trophoblast proliferation on inner cell mass derivatives persists into later gestation and, although its basis remains obscure, may be an important factor in ensuring harmony of development between trophoblast and embryo.

Differentiation of cells within the inner cell mass

Inner cell masses can be isolated from very early blastocysts by immuno-surgery, and are able to re-form into small blastocysts in culture by regenerating an outer layer of trophectoderm cells. However, if isolated somewhat later, they respond to culture by forming an outer layer of endoderm rather than trophectoderm. Indeed, Janet Rossant found that even if more mature inner cell masses were aggregated with morulae, they could not be persuaded to contribute to the trophoblast of the resulting chimaeras. It seems, therefore, that inner cell mass and trophectoderm cells are restricted to mutually exclusive pathways of differentiation by the expanded blastocyst stage.

Injection of genetically labelled inner cell masses into either blastocysts (Fig. 6.2) or, as discussed earlier, trophectodermal vesicles, can be used to determine the fate of this tissue as a whole. However, such experiments cannot tell us anything about the status of individual inner cell mass cells. It might be that each of these cells can contribute either to all inner cell mass derivatives or only to particular ones. Since inner cell mass cells look alike at $3\frac{1}{2}$ days, this problem can only be resolved by studying the fate of single cells. Mitotic descendants of a single cell are known collectively as a clone, so that this approach is termed clonal analysis of development.

Approximately 20 per cent of single cells from disaggregated $3\frac{1}{2}$-day inner cell masses yield detectable chimaerism following injection into blastocysts. The frequency of chimaeras can exceed 50 per cent if daughter cell pairs are injected. The resulting clones are often very large indeed, and can extend to all tissues of the conceptus that are known from blastocyst 'reconstitution' experiments to originate from the inner cell mass (Table 6.2). Evidently, inner cell mass cells are not restricted in their developmental potential at $3\frac{1}{2}$ days, except with regard to trophectodermal differentiation.

The situation is very different by $4\frac{1}{2}$ days of development when the inner cell mass clearly consists of two tissues (Fig. 6.3), whose cells can be distinguished following its dissociation. Very clear-cut reciprocal patterns of chimaerism are seen in conceptuses that have developed from blastocysts injected with primitive endoderm versus primitive ectoderm cells. Primitive endodermal clones are confined to one or both layers of the extraembryonic endoderm only (Table 6.2). Primitive ectodermal clones can, in contrast, extend throughout the fetus and embrace the amnion, allantois and mesodermal layer of the yolk sac as well (Table 6.2). In addition to this impressive array of cell types, individual primitive ectoderm clones have also been found by test-breeding adult chimaeras to form functional eggs or spermatozoa. Hence, the primitive ectoderm seems to give rise to the entire soma of the mouse, including its endodermal organs, and also to the germ cell line, and certain extraembryonic tissues. Primitive endoderm, on the other hand, appears like trophectoderm to play a purely extra-embryonic role. Primitive ectoderm and primitive endoderm thus behave like epiblast and hypoblast, respectively, in the chick embryo, and are consequently often referred to by these terms.

The data on cell lineage that have been obtained so far by following the fate of genetically marked cells or tissue within the embryo are summarized diagrammatically in Fig. 6.7. There are problems in extending this analysis to later stages of development because embryos cannot be removed and returned to the uterus once they have implanted. Janet Rossant and her colleagues found that although cells from trophectodermal and primitive endodermal derivatives of post-implantation embryos could colonize the blastocyst, those of primitive ectodermal origin consistently failed to do so. Hence, current knowledge of the fate of various regions of the post-implantation embryo is based largely on analysing their differentiation in extrauterine sites such as the kidney or testis. At best, this approach provides a very crude indication of their normal role *in situ*. However, such ectopic grafts can give rise to transplantable tumours called teratocarcinomas which, as discussed later, have proved a valuable alternative to the embryo for studying a number of aspects of differentiation.

Very recently, Rosa Beddington and Andrew Copp in Oxford have developed a method for exchanging grafts of small pieces of tissue between post-implantation embryos following their recovery from the uterus. By exploiting culture techniques pioneered by Denis New in Cambridge, they

Fig. 6.7. Scheme of cell lineages in the early mouse embryo based on transplantation of genetically marked cells and tissue into blastocysts. The approximate number of cells in certain tissues at the time they become restricted in developmental potential is indicated in parenthesis.

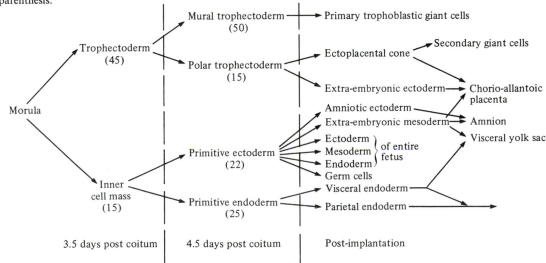

were able to show that most grafted embryos can continue to develop normally *in vitro* for 36 hours or more at a rate comparable to those remaining *in utero*. Radioactive labelling of the donor cells prior to grafting enables their fate to be examined with precision after the host embryos are sectioned and stained. Interesting results are beginning to emerge from this work, but the duration of the experiments is clearly limited by the time for which normal growth of the host embryos can be sustained in culture.

Blastocyst biopsy – practical applications

The mouse blastocyst is evidently very resilient to microsurgery, and rabbit and sheep blastocysts can also develop normally after the removal of some trophoblast. This raises the possibility of studying the excised tissue to obtain information about the embryo. The type of information that can be gained will depend both on the amount of tissue available and also on the genes that it expresses. Such an approach offers the important advantage over amniocentesis that implantation has not yet taken place, and embryos may be selected for transfer according to the results obtained.

The most obvious way that this technique might be applied is in the control of the sex of offspring. Hitherto all attempts to do this have been directed without success towards separating spermatozoa bearing X and Y chromosomes. The author, together with Robert Edwards in Cambridge, was able to sex live rabbit blastocysts by a microsurgical technique. Each blastocyst was held by suction on the tip of a glass pipette, and a column of trophectoderm was drawn out by suction from the opposite side. The exposed trophectoderm was excised with a pair of very fine scissors (Fig. 6.8), and the sample was stained and scored for sex chromatin. The 'sexed' blastocysts were transferred to recipient females after they had recovered in culture. Eighteen developed to term, and the sex of all was found to have been predicted correctly at the blastocyst stage.

This particular procedure cannot be applied to the blastocyst in all mammals – in most species studied sex chromatin does not appear until after implantation, and there are fewer cells in the entire blastocyst of the rat, mouse, hamster, guinea pig or human than in the average biopsy specimen from the much larger rabbit blastocyst. Nevertheless, Charles Epstein and his colleagues in San Francisco recently demonstrated the feasibility of sexing mouse blastocysts by a different approach. They separated the blastomeres of 2-cell embryos and cultured them in isolation. In 71 per cent of cases both blastomeres developed into half-sized blastocysts, one of which was used to prepare chromosome spreads for sexing its partner. Satisfactory chromosome preparations were obtained in 47 per cent of cases so that one-third of the initial embryos produced living half-blastocysts of known sex. However, the potential for further development of the latter was not tested in this study, which was undertaken for a different reason.

If developed in conjunction with techniques of superovulation and non-surgical transfer, a procedure based on typing blastocysts could assume practical as well as scientific value. For example, certain serious hereditary diseases in man such as haemophilia and the Duchenne type of muscular dystrophy are due to recessive X-linked genes, and thus afflict sons much more frequently than daughters. Selection of daughters in women who are known to carry the defective gene could avert the birth of affected individuals in perhaps a more acceptable way than the present methods of amniocentesis and abortion. The social and ethical implications of the selection of embryos are discussed in Book 5.

Teratocarcinomas

These tumours can develop spontaneously from male germ cells in one inbred strain of mice, and from female germ cells that embark on parthenogenetic development in the ovary in another. Teratocarcinomas can also be produced experimentally in several inbred strains of mice by transplanting pre- or early post-implantation embryos ectopically. The resulting growths contain a wide variety of differentiated tissues, including nerve, gut, muscle and cartilage, which are jumbled together in a chaotic fashion (Fig. 6.9). Many are designated benign teratomas because they stop growing after a while and begin to regress. The remainder, however, continue to grow and will eventually kill the host. Unlike teratomas, these

Fig. 6.8. Removal of piece of trophectoderm for sexing the living rabbit blastocyst. The blastocyst is held by gentle suction (*left*) while a column of trophectoderm is withdrawn from the opposite side. The exposed trophectoderm is then excised with a pair of fine scissors. (From R. L. Gardner and R. G. Edwards, *Nature, Lond.* **218**, 346 (1968).)

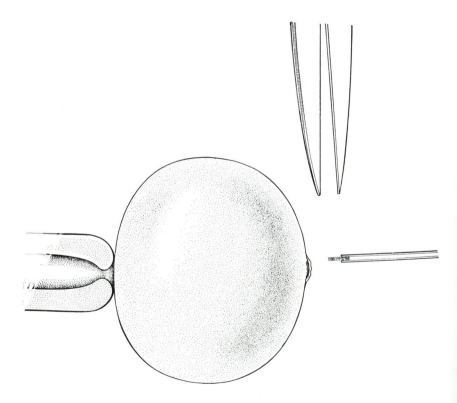

malignant teratocarcinomas can be maintained indefinitely provided that they are transplanted at intervals to adults of the same inbred strain. The crucial difference between a teratoma and a teratocarcinoma is that the latter contains a self-renewing stock of stem cells called embryonal carcinoma cells.

When transplanted subcutaneously, teratocarcinomas remain as solid tumours in which the embryonal carcinoma cells are widely scattered in small nests. However, by inserting pieces of the solid tumour into the peritoneal cavity, Barry Pierce and his colleagues at the University of Colorado found that they can sometimes be made to form vast numbers of discrete spherical or oblong bodies. These structures are termed embryoid bodies because they clearly resemble the egg cylinder of the early post-implantation embryo. In their simplest form, embryoid bodies consist of a solid core of embryonal carcinoma cells surrounded by a rind of endoderm-like cells. They can also become cystic, in which case a variety of differentiated cell types can be distinguished inside the endoderm layer. The same group of investigators have grafted single embryonal carcinoma cells into adult mice, exploiting embryoid bodies as a concentrated source

Fig. 6.9. Section of part of a teratocarcinoma to show some of the differentiated tissues that such a tumour is capable of producing. BM, bone matrix; C, cartilage; CT, connective tissue; E, epithelium; KE, keratinizing epithelium; KS, keratinized squames produced by keratinizing epithelium; NE, nucleated erythrocytes. courtesy of Dr Christopher Graham.)
(Photograph by courtesy of Dr Christopher Graham.)

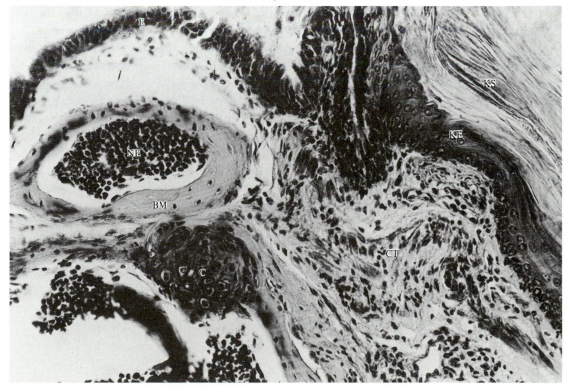

of these cells. The majority of the resulting tumours were indistinguishable from those produced by grafting whole pieces of tumour, both in terms of their transplantability and the range of differentiated cell types that they formed. Hence the embryonal carcinoma cell is clearly the teratocarcinoma stem cell and is multipotential.

Embryonal carcinoma cells have also been established in culture both from embryoid bodies and solid tumours. Indeed, many cloned lines of these cells are now available, some of which have been maintained *in vitro* for considerable periods. It is, in fact, precisely because teratocarcinomas are a source of pluripotent cells which can be adapted so readily to culture that they have attracted so much attention. By studying embryonal carcinoma cells in culture, specific steps in differentiation can be analysed with a precision that is unattainable in the embryo. Furthermore, large scale cultures can be used to provide homogeneous populations of cells in the relatively large quantities necessary for certain biochemical and immunological investigations. Nevertheless, one might legitimately question the use of malignant tumour cells to study normal development. This is why it is of particular interest to examine the consequences of re-introducing them into the normal embryo.

Ralph Brinster in Philadelphia was the first person to inject embryonal carcinoma cells into blastocysts, the donor cells being obtained from a tumour passaged *in vivo*. One of the resulting offspring showed an unequivocal contribution of the donor cells to the melanocytes populating the hair follicles in the coat. Subsequently, Beatrice Mintz and her colleagues, also working in Philadelphia, confirmed and extended this work. They demonstrated that embryonal carcinoma cells from in-vivo tumours could participate in the formation of normal chimaeric mice by contributing to many different tissues, and they even gave rise to functional gametes in three instances. This is dramatic testimony to the capacity of these tumour cells for normal differentiation, especially since one of the tumours had been passaged for 8 years before it was used to provide cells for blastocyst injection. Independently, Ginny Papaioannou and her colleagues in Oxford showed that embryonal carcinoma cells that had been maintained in culture could also participate in normal development following injection into blastocysts. However, the hope that the cultured cells might also form functional gametes has not been realized so far. If cultured tumour cells could be made to behave this way, this would be particularly interesting, since new mutations can readily be generated and selected in cultured cells. Manipulating embryonal carcinoma cells in this way would obviously provide a means of introducing interesting new mutations into mouse stocks if the cells formed functional eggs or spermatozoa following introduction into the blastocyst.

Other uses of experimental chimaeras

We have seen how experimental chimaeras can be used to study the fate of both normal cells of the early embryo and their malignant counterparts. This is just one of their many applications. Indeed, so varied are the ways in which these fascinating organisms have been exploited that it is impossible to discuss all of them here. Instead, a few examples have been chosen that we hope will give some idea of the range of uses to which they can be put.

Differentiation of skeletal muscle in vivo

Skeletal muscle fibres are long spindle-shaped structures containing a central bundle of contractile apparatus and a peripheral array of nuclei. These fibres develop in culture by the fusion of several separate myoblasts (precursor cells), each of which contains a single nucleus, rather than from a single myoblast, by repeated nuclear division without cytoplasmic cleavage (Fig. 6.10). To decide whether differentiation of skeletal muscle proceeds similarly *in vivo*, Beatrice Mintz and Wilber Baker in Philadelphia carried out the following elegant experiment. They aggregated pairs of mouse eggs that were homozygous for different forms (allelic variants) of the enzyme isocitrate dehydrogenase. The two forms of the enzyme differ in electrophoretic mobility, type *a* migrating slowly and type *b* more rapidly towards the anode. Mice that are F_1 hybrids show an intermediate *ab* band, in addition to the pure *aa* and *bb* forms (Fig. 6.10). However, hybrid enzyme cannot be formed by mixing the two pure types *in vitro*. Various tissues from chimaeric offspring were prepared for electrophoresis. The results were unequivocal. Cardiac muscle, liver, spleen and other tissues showed only the *aa* and *bb* bands, whereas skeletal muscle was unique in showing the *ab* form as well. This result affirms that cell fusion occurs in the development of skeletal muscle in the body just as was found in culture. The possibility that trophoblastic giant cells are also formed by cell fusion has been investigated in chimaeras by using another enzyme that behaves like isocitrate dehydrogenase. The absence of hybrid molecules in those cells supports other evidence suggesting that they are, in fact, formed by repeated replication of DNA without concomitant cell division.

Sexual differentiation

When pairs of cleaving embryos are aggregated to make chimaeras, those of opposite sex are apt to be combined as often as those of like sex. Hence we might expect half the overt chimaeras to be XX/XY and therefore very relevant for the investigation of sexual differentiation. Surprisingly, hermaphrodites or intersexes are rare in all aggregation chimaeras produced so far. How is this explained? Two obvious alternatives are that either XX/XY chimaeras mainly die *in utero* or that they are born but pose as respectable males or females. Support for the latter notion comes from examining the sex ratio of chimaeric offspring, which shows a significant

Fig. 6.10. Division (*left*) and fusion (*right*) models to explain the development of skeletal muscle in mouse chimaeras, together with the expected electrophoretic patterns of isocitrate dehydrogenase ('muscle zymograms'). The fusion model was found to be correct. (After B. Mintz and W. W. Baker, *Proc. Natl. Acad. Sci.*, *USA*, **58**, 592 (1967).)

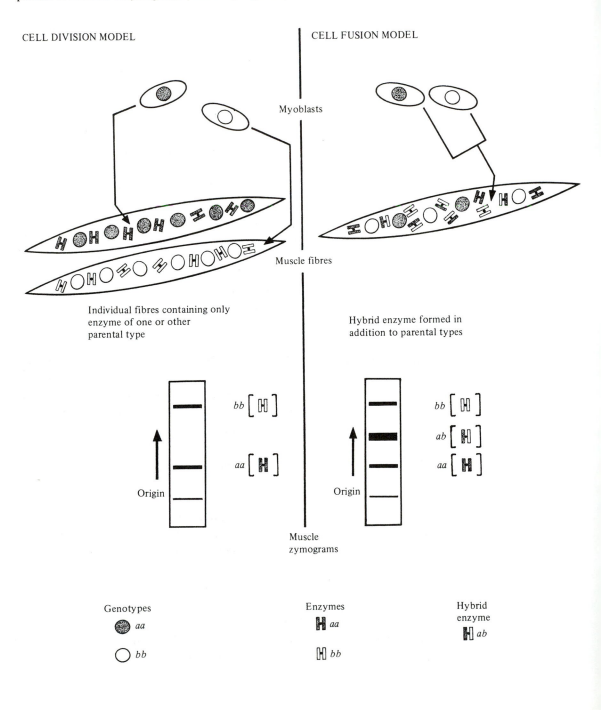

excess of males in most studies. In view of the dominant role of the Y chromosome, Krystof Tarkowski in Warsaw suggested that XX/XY mosaicism may lead to a male phenotype. Examination of metaphase chromosomes from bone marrow confirmed that some fertile males were indeed of this type. However, Beatrice Mintz found a normal sex ratio in another extensive series of chimaeras, and identified XX/XY females as well as males. The situation was clarified by Richard Mullen and Wesley Whitten working at the Jackson Laboratory in Maine. They showed that either an excess of phenotypic males or a normal sex ratio could be obtained, depending on the strains from which the combined embryos were derived. Some combinations yielded 'balanced' chimaeras exhibiting roughly equal contributions from the two constituent embryos: these included more phenotypic males than females. In other 'unbalanced' combinations the sex ratio was normal because the sex of the chimaeras was typically that of the embryo of the dominant strain.

Beatrice Mintz further demonstrated that XX/XY male chimaeras may be fertile with histologically normal testes, even though the latter contain a high proportion of XX cells. This raises the interesting question of the origin of the functional germ cells in these animals: are they likewise a mixture of genetically male and female cells? Chimaeras obtained after union of embryos of the same sex often show chimaerism in the germ line as in other tissues, and two genetically distinct types of gametes may be produced by one gonad.

The results of test-breeding fertile XX/XY male chimaeras are unambiguous. Their offspring have a normal sex ratio and are all of one type, which invariably corresponds with the XY component of the chimaera. Functional reversal of differentiation of XX germ cells does not occur in the mouse, nor apparently in several other mammals in which autosomal genes can cause testicular development in genetic females (see Chapter 3). What happens to XX germ cells in XX/XY male chimaeras? Ewa Mystkowska and Krystof Tarkowski found, on examining the testes of late fetuses, germ cells in meiotic prophase side by side with normal prespermatogonia. Anne McLaren and her colleagues obtained evidence that these were XX germ cells entering meiosis on schedule, in defiance of the testicular environment around them. Such oocytes disappear soon after birth.

As noted earlier, XX/XY female chimaeras are in general less common than males, and have so far been studied much less extensively. Recently, Ted Evans and his colleagues in Oxford found a mature XY oocyte in one female chimaera. However, it has yet to be established whether such an oocyte can be fertilized and develop into a normal mouse.

Cell deployment during development
Obviously, the lineage studies discussed earlier depend on the use of a marker gene that is expressed in all cells of the embryo and adult. Allelic

variants of the gene for glucose phosphate isomerase provide the single ubiquitous cell marker available at present. Unfortunately, the two forms of this enzyme can only be distinguished by electrophoresis of tissue homogenates. Hence, while this marker indicates the proportions of cells of each genotype in a tissue, it can tell us very little about their spatial arrangement. Other genes can be used as markers to visualize patterns of chimaerism *in situ*, but these are expressed in one or, at most, a few specific types of cell. The value of an *in situ* marker is illustrated most graphically when the constituent cells of chimaeras carry genes for different coat colour such as albinism versus pigmentation. Patterns are then clearly visible on the surface of the living animal. Pepper and salt markings, which would be due to complete intermixing of albino and pigmented hairs, are never seen. Instead, more or less discrete patches of one or other colour are the rule. These tend to be arranged transversely and are often asymmetrical, thereby revealing a discontinuity along the dorsal midline (Fig. 6.11). Such patches arise through clonal proliferation of pigment precursor cells that have migrated laterally from the neural crest in the embryo. The patches will obviously vary in size depending on whether adjacent clones are of the same or opposite type.

Fig. 6.11. Mouse chimaeras obtained by transferring a single inner cell mass cell from a pigmented to an albino blastocyst (*left*), or by transferring an entire albino inner cell mass to a pigmented blastocyst (*right*). Pigmented and albino areas are organized into discrete patches on either side of the midline.

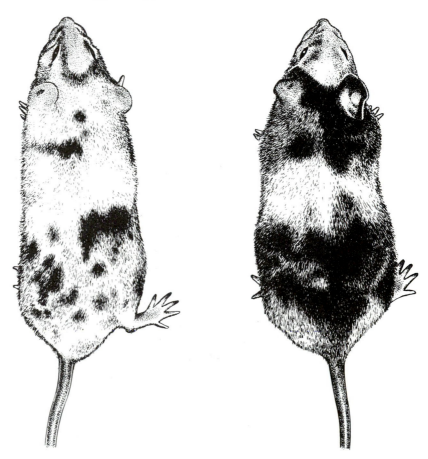

Clearly, much can be learnt about cell deployment during development from examining patterns of chimaerism *in situ*. It is for this reason that considerable effort is being made to find a marker that would enable all tissues to be investigated in this way.

Investigation of X-inactivation

In female mammals one X chromosome is inactivated early in development and appears in the nucleus as the sex chromatin or Barr body. The production of mouse chimaeras by injecting one cell into each host blastocyst provides a means of determining the time of X-inactivation in this species. The principle of the test is to so arrange mating that donor embryos, if female, will have two distinct coat colour genes controlled by their two X chromosomes. A single donor cell is then injected into each of a series of host blastocysts that carry a gene specifying a third colour. If the female donor cell has already undergone inactivation at the time of transfer, the resulting chimaera will show only one of the two donor colours together with the host colour, whereas if no inactivation has occurred a three-coloured mouse will be expected.

Preliminary results of experiments performed in collaboration with Mary Lyon of Harwell indicate that X-inactivation has not taken place in the primitive ectoderm of the $4\frac{1}{2}$-day implanting blastocysts, though it probably does so a few cell divisions later. Astonishingly, as much as half the resulting mouse can be formed by progeny of the single injected cell.

Rescue of lethal mutants

The conventional way of finding out whether two mutations are alleles at the same chromosomal locus is to do a complementation test by mating males carrying one mutation with females carrying the other. If the resulting offspring are normal, it means that the two mutations are able to complement each other, and must therefore involve different loci. If, on the other hand, the offspring are mutant, the mutations must have occurred at the same locus.

There are two mutations in the mouse called 'jimpy' and 'myelin-synthesis deficiency', respectively, which had been found to occupy roughly the same location on the X chromosome. However, it was not possible to determine whether they were alleles by a direct complementation test in this instance because male offspring carrying either mutation on their single X chromosome die from impaired function of the brain before they can reproduce. Heterozygous females can develop into fertile adults presumably because, following random X-chromosome inactivation during development (see Chapter 3), they have sufficient brain cells in which the normal gene is active. Eva Eicher and Peter Hoppe at the Jackson Laboratory reasoned that mutant males might be able to reproduce if they could be provided with enough non-mutant brain cells, so they aggregated embryos from matings designed to produce 'jimpy' males with non-mutant embryos from an inbred stock. From this they were able to produce an unbalanced male chimaera that was predominantly 'jimpy', but survived to reproduce despite showing transient effects of the mutation. This male proved to have a 'jimpy' germ cell line and produce mutant female as well as male offspring when mated with females that were heterozygous for myelin-synthesis deficiency. Hence it was concluded that the latter mutation was indeed an allele of 'jimpy'. Certain mutations which act as autonomous lethals, either in all cells of the early embryo or specifically in the germ cell line obviously cannot be rescued in this way.

Transplantation of nuclei

The technique of injecting isolated nuclei into enucleated eggs was devised originally for the relatively large amphibian egg by Robert Briggs and Thomas King working in Philadelphia. It has proved to be a valuable way of studying nuclear changes and relations with the cytoplasm during development. However, application of nuclear transplantation to the much smaller mammalian egg has defied solution for nearly a decade. The chief problem has been bursting of the egg following penetration with a pipette that is large enough to accommodate donor nuclei. However, very recently, Karl Illmensee and Peter Hoppe found that nearly 40 per cent of eggs remained intact if the operation was carried out in medium containing cytochalasin B. It is not clear how this mould metabolite protects the egg, but since it interferes with the assembly of microfilaments it may do so by relaxing the cortical cytoskeleton. Trophectoderm or inner cell mass cells from 4-day blastocysts provided donor nuclei which were injected into fertilized eggs that had reached the pronuclear stage. Injection pipettes were designed so as to rupture the membrane of donor cells without damaging their nuclei. Eggs were held by suction on the tip of a second pipette while the donor nucleus was injected. Finally, both host pronuclei were sucked into the injection pipette before it was withdrawn (Fig. 6.12).

Approximately half of the eggs remaining intact after operation started to divide in both the inner cell mass and trophectoderm nuclear transplantations. Half of the eggs embarking on cleavage that received cell nuclei from the inner cell mass reached the morula or blastocyst stage, whereas less than one-fifth of those receiving trophectoderm nuclei did so. Three young were born following transfer of 16 embryos from the inner cell mass nuclear transplant series to uterine foster-mothers. All three displayed the genetic characteristics of the nuclear donor embryos. Two of them grew into fertile adults, the third dying inexplicably at 7 weeks.

The extraordinary power of this nuclear cloning technique is the conclusiveness of positive results; it clearly demonstrates that the nuclei of inner cell mass cells are totipotent. However, negative results cannot automatically be taken to mean that the genome has undergone irreversible change during differentiation. They might, for example, reflect differences in susceptibility of nuclei of different cells to damage during manipulation.

Fortunately, sensitive biochemical techniques are now available that can be used to determine whether particular sequences of the DNA are chemically modified, deleted or transposed.

The past decade has witnessed an explosion of interest in the early mammalian embryo as an object for experimental investigation. The remarkable resilience of the cleaving egg and blastocyst may be exploited in several different ways to gain information on development and exert some degree of control. Experimental chimaeras are probably the most valuable products of these studies so far. Their potential for the analysis of development as well as for research on cancer and problems of

Fig. 6.12. Nuclear transplantation into the fertilized mouse egg at the pronuclear stage. (*a*) Before injection of the nucleus, and (*b*) after injection of the nucleus and during removal of host pronuclei. (By courtesy of Karl Illmensee.)

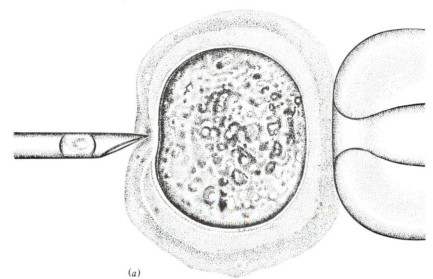

(*a*)

(*b*)

immunology is considerable. Further rewards are just around the corner. The repair of genetically faulty blastocysts by the inoculation of cells from normal ones may soon be possible. Nuclear transplantation could have a tremendous impact by allowing the immediate and precise selection of genotype. Prospects in this field look bright indeed.

Suggested further reading

Normal mammalian muscle differentiation and the gene control of isocitrate dehydrogenase synthesis. B. Mintz and W. W. Baker. *Proceedings of the National Academy of Sciences, USA*, **58**, 592–598 (1967).

Properties of extra-embryonic ectoderm isolated from postimplantation mouse embryos. J. Rossant and L. Ofer. *Journal of Embryology and Experimental Morphology*, **39**, 183–94 (1977).

Investigation of the fate of 4.5 day *post-coitum* mouse inner cell mass cells by blastocyst injection. R. L. Gardner and J. Rossant. *Journal of Embryology and Experimental Morphology*, **52**, 141–52 (1979).

Nuclear transplantation in *Mus musculus:* developmental potential of nuclei from preimplantation embryos. K. Illmensee and P. Hoppe. *Cell*, **23**, 9–18 (1981).

Transplanted nuclei and cell differentiation. J. B. Gurdon. *Scientific American*, **219**, December, 24–35 (1968).

Choosing sex before birth. R. G. Edwards and R. L. Gardner. *New Scientist*, **38**, 218–220 (1968).

Teratocarcinomas and mammalian embryogenesis. G. R. Martin. *Science*, **209**, 768–776 (1980).

Manipulations on the blastocyst. R. L. Gardner. In *Advances in the Biosciences 6*. Schering Symposium on Intrinsic and Extrinsic Factors in Early Mammalian Development, Venice, 1970. Ed. G. Raspé. Pergamon Press; Vieweg (1971).

The relationship between cell lineage and differentiation in the early mouse embryo. R. L. Gardner. In *Genetic Mosaics and Cell Differentiation: Results and Problems in Cell Differentiation*, vol. 9. Ed. W. Gehring. Springer-Verlag; Berlin (1978).

Sexual differentiation in mammalian chimaeras and mosaics. A. McLaren. In *Genetic Mosaics and Cell Differentiation: Results and Problems in Cell Differentiation*, vol. 9. Ed. W. Gehring, Springer-Verlag; Berlin (1978).

Mammalian Chimaeras. A. McLaren. Cambridge University Press; London (1976).

INDEX

Numbers in italics indicate pages on which Figures. appear; those printed **bold** *carry Tables.*

abortion
  spontaneous, *144*, **145**, 147, 148, 149, *150*, *151*, 152, 154–6, 158
  see also pregnancy loss
achondroplasia, 157
adrenogenital syndrome, 102
agouti, gene functional in 2- to 4-cell mouse embryo, 13
albino locus deletion *c25H*, active in 2- to 4-cell mouse embryo, 13
allantois
  fluid contains 'purple protein' in pig, 43
  formation, 46, *47*, *49*
amnion, formation of, 46, *47*, *49*
aneuploidy
  in mouse and rabbit embryos, 14
  see also under chromosome
antelope, form of placenta, *51*, **52**
antibody, transmission from mother to fetus, 63
ape
  form of placenta, **52**
  type of implantation, *63*
  see also chimpanzee
armadillo
  blastocyst remains in suspended development after 'hatching', 29
  ovariectomy causes implantation of diapausing embryo, 38
armadillo, nine-banded, *Dasypus novemcinctus* identical quadruplets, *20*, 21
  ovariectomy induces

implantation or loss of embryo, 39
armadillo, twelve-banded, eight monozygotic young, 21
Artibeus jamaicensis, time of embryonic diapause related to rains, 37
Artiodactyla
  blastocysts elongate greatly before attachment, and implantation is on endometrial caruncles, *30*
  preimplantation changes in embryo, *2*

badger, European, *Meles meles*
  activation of blastocyst, 34, *35*
  duration of embryonic diapause, 27
bandicoot, develops chorioalantoic placenta, 57
bandicoot, *Isoodon obesulus*, sex chromosomal status of somatic and germ cells differ, 85, *86*
Barr body, *80*, 81
bat
  delayed fertilization and retarded development, 34
  discoid placenta, 62
  preimplantation changes in embryo, *2*
  X chromosome translocated onto autosome, 78
bat, African fruit, *Eidolon helvum*, rain associated with reactivation of diapaused blastocyst, 37

bat, American fruit, *Artibeus jamaicensis*, time of embryonic diapause related to rains, 37
bat, bent-winged, *Miniopterus schreibersii*
  blastocyst in diapause, *26*
  implantation of diapausal blastocyst after hibernation, 37
bear
  characteristics of embryonic diapause, **28**
  form of placenta, *51*, **52**, 62
birth
  causes of defects of, 142–6, 156, 157
  circadian rhythms of, 140, 141
  preparation for, 133–5
birth defects
  achondroplasia, 157
  cystic fibrosis, 157
  Down's syndrome, mongolism, trisomy-21, 148, *149*
  effects of cytomegalovirus, 157
  effects of thalidomide, 6-mercaptopurine, dilantin, 157
  general aspects, 142, *143*, *144*, **145**, 145, 146, *149*, *150*, *151*, *152*, **153**, 156, 157
  haemophilia, 156, 157
  muscular dystrophy, 156, 157
  Rubella syndrome, 157
  Tay–Sachs disease, 157
blastocoele
  description, *2*, 3, 4

blastocoele (*cont.*)
　passage of fluid into, due
　　to ion pump, 18
　relation to future yolk
　　sac, 46, *47*, *49*
blastocyst
　activation, 34–7
　contraction and
　　expansion in culture,
　　17, 18
　differentiation of cells
　　within inner cell mass,
　　164, 167–9
　fluid composition, 4, 5
　formation, *2*, 3–7, *4*, *5*, *6*
　'hatching', *2*, 27, 29
　ion pump in, 18
　'reconstituted',
　　development of, 163,
　　165, *166*, 167
　spacing out in uterus, 31
　structure in several
　　mammals, *28*
　transfer between sheep, *8*,
　　9, 10
blastokinin, 17, 42

camel, with diffuse
　placenta, 52
carbon dioxide output by
　embryo, *10*, 11
caruncle, development and
　function, 30, 59, 60
cat
　form of placenta, *51*, 62
　times of stages of
　　gestation, and total
　　time, **22**
　tortoiseshell: visible
　　evidence of X
　　inactivation, 79, 80;
　　males either XXY in
　　karyotype or
　　chimaeras, 80
cattle
　elongation of blastocyst,
　　2, 3
　form of placenta, *51*, **52**
　growth of 'hatched'
　　embryos, *30*
　placental anastomosis, 77
　some asynchrony
　　between embryo and
　　uterus tolerated, 42
　transuterine (internal)
　　migration of embryos
　　rare, 24
　twinning, placental

anastomosis and
　freemartinism, 19, 76, 77
see also cow
chimaera
　experimental, 7, 159–61,
　　173, *174*, 175, *176*,
　　177, 178
　legendary, *159*
　spontaneous, 80, 161
chimpanzee, interstitial
　implantation, 30
chorion, formation, 46, *47*,
　49
chorionic gonadotrophin
　described in man, rat,
　　mouse, rabbit, pig and
　　horse, 39
　equine, 52, 56
chromosome
　abnormalities in human
　　spontaneous
　　abortions, *144*, **145**
　monosomy, 14, **145**, 147,
　　148, 149, *150*
　non-disjunction, 147, *148*
　single-gene defects, 144,
　　145, 155–7
　tetraploidy, 9, **145**, 150
　triploidy, 7, **145**, 150,
　　151, 152
　trisomy, *14*, *15*, *143*, **145**,
　　147, *148*, *149*, 156
　trisomy-13, -18, -21, 156
　X-inactivation, 16, 79,
　　80, 177, 178
congenital adrenal
　hyperplasia, 102
cotyledon, in ungulate
　placenta, *51*, 59, *60*
cow
　blood oestrogen and
　　progesterone levels
　　before parturition, *137*
　'bulling' behaviour with
　　cystic ovary, 108, 109
　embryo in
　　allantochorionic sac, *60*
　source of progesterone in
　　late pregnancy and
　　presence of placental
　　17α-hydroxylase, **138**
　times of stages of
　　gestation, and total
　　time, **22**
　see also cattle
Crocuta crocuta, male and
　female external
　genitalia, *105*, 106

culture (cultivation) *in vitro*
　of preimplantation
　　embryo, 7, 16–18
　of postimplantation
　　fetus, 18
cyclic adenosine
　monophosphate
　　(cAMP), changes in
　　uterus after
　　stimulation of decidual
　　cell response, *33*, 34
cystic fibrosis, 157
cytomegalovirus, effects of,
　157

Dasypus novemcinctus, see
　nine-banded armadillo
decidua
　cell reaction, 32–4, *33*
　types, 65, 68
deer
　control of embryonic
　　diapause, *41*, 42
　development suspended
　　after blastocyst
　　'hatching', 29
　female with male
　　behaviour after
　　testosterone implant,
　　107, *108*, 109
　form of placenta, **52**
deoxyribonucleic acid
　(DNA)
　embryo stage at first
　　transcription, 13
　synthesis during
　　cleavage, 3
development, manipulation
　of
　analysis of differentiation
　　of cells in inner cell
　　mass, **164**, 167,
　　168, 169
　chimaera: experimental,
　　7, 159–61, 173,
　　174, 175, *176*, 177,
　　178; spontaneous, 80,
　　161
　comparison of isolated
　　trophectoderm and
　　inner cell mass tissue,
　　160, 162, *163*, **164**,
　　165, *166*, 167
　destruction of one or
　　more blastomeres
　　experimentally, 9
　development of
　　'reconstituted'

development (*cont.*)
 blastocysts, 163, 165,
  166, 167
 experimental
  parthenogenetic
  activation, 7
 investigation of
  X-inactivation,177,178
 teratocarcinoma,
  production and study,
  170, *171*, 172
 transplantation of nuclei,
  178, *179*, 180
 use of blastocyst biopsy,
  169, *170*
 see also under embryo
diapause
 blastocyst of bat
  Miniopterus
  schreibersii, 26; of
  wallaby *Macropus*
  eugenii, 26
 characteristics of, in
  different mammals, **28**,
  34
 DNA polymerase in
  normal and diapause
  mouse embryos, *38*
 duration, 27
 endocrine control, 37–42
 environmental control,
  34, *35*, *36*, 37
Dicrostonyx torquatus, *see*
  lemming, Arctic
  varying
differentiation in embryo,
 4, 5
 see also under embryo
dilantin, effects of, 157
DNA polymerase, activity
 in normal and
 diapausing mouse
 embryo, 38
dog
 blastocyst expands
  before implantation, 29
 form of placenta, *51*, **52**,
  62
dolphin, form of placenta,
 51, **52**
Down's syndrome, *143*,
 144, 148, *149*
dugong, form of placenta,
 52, 62

echidna
 nutrients in egg derive
  mainly from uterus, 46

structure of egg and
 early development, 48
Eidolon helvum, rain
 associated with
 reactivation of
 diapaused blastocyst,
 37
elephant, African, early
 and late stages of
 placenta development,
 61, 62
Ellobius lutescens, lacks Y
 chromosome, 78
Embden-Meyerhof
 pathway in embryo
 metabolism, 11
embryo
 biochemical studies on,
  10, **11**, 12, 13
 block to cleavage *in vitro*
  at 2–cell stage, 17
 cell lineages in early
  mouse embryo, **164**,
  167, *168*, 169
 chromosome aberrations
  produced by
  antimetabolites, 14
 cleavage, 1–3, *21*
 compaction, 3
 culture *in vitro*, 16–18
 destruction of one or
  more blastomeres
  experimentally, 9
 DNA polymerase and
  number of cells in
  normal and diapause,
  38
 ectopic implantation, 42,
  45
 energy sources for
  development, **11**
 experimental
  manipulation, 7–10, *8*,
  9; *see also*
  manipulation of
  development
 experimental
  parthenogenetic
  activation, 7
 external migration, 24
 gene action, 13–16
 internal migration, 23
 inter-species hybrids
  (rabbit–hare) with
  normal cleavage, 13,
  14
 preimplantation changes,
  2

preimplantation
 embryo–maternal
 interactions, *40*
production of
 gonadotrophins in
 peri-implantation
 period, 39
quadruplets, 18, *20*, 21
rate of transport, *21*
stages of development,
 21
times of entry into
 uterus, and
 implantation, *22*
transfer to recipient, 7, *8*,
 9, 10, *23*, 24
twins: spontaneous, 18,
 19; experimental
 production of, *8*, *9*, 10
 see also blastocyst,
 diapause
embryonic disc, in pig
 blastocyst, *4*
endometrial caruncle,
 implantation site in
 artiodactyl, 30, 59, 60
endometrial cup
 in hinney, 55
 in horse, 52, *55*, 56,
 in mule, 55
eunuch, 99, *100*
euro (hill kangaroo)
 Macropus robustus,
 effect of nutrition on
 embryonic diapause,
 37

fat, brown and white
 distribution of and
 function as energy
 stores in fetus, *134*,
 135
felid, form of placenta, *51*,
 52
ferret
 blastocyst expansion
  before implantation,
  29
 form of placenta, *51*, 62
 times of stages of
  gestation, and total
  time, **22**
fertilization, 'delayed', in
 bats, 34
fetus
 adrenal glands, *129*, 130
 carbohydrate and fat
  reserves, *134*, 135

fetus (*cont.*)
  cardiovascular system, *121*, 122
  cell growth, 118
  circadian rhythms of birth, 140, 141
  control of parturition, 135–41, *137*, *138*, **138**, *139*
  cortisol promotes oestrogen synthesis in placenta, 138
  culture *in vitro*, 18
  diabetes, feto-maternal relations, *128*, 129
  diet, 115
  effects of fetal hormones on mother, 126
  endocrine system, 125–30
  'facilitated diffusion' of glucose from mother to fetus, 116
  fetal genotype and parturition, 135
  feto-placental unit, 125, 126
  functions, 120–33, *123*, *128*, *129*; fetal hypothalamus, 127; fetal pituitary gland, 127; fetal thyroid, 127, 128
  gastro-intestinal system, 122, 123
  gonads, 129
  growth, 114–20, *117*, *119*
  immunological aspects, *44*, 45, 130, 131
  initiation of lactation, 135
  little capacity for gluconeogenesis, 115
  loss, 142–6, *143*, *144*, **145**; *see also* pregnancy loss
  maternal hormones normally lack effect on, 126
  mechanism for fetal control of labour, 139; *see also* parturition
  nervous system, 131–3
  pancreas, 128, 129
  preparation for birth, 133–5
  prostaglandins and parturition, 135–40, *137*, *139*
  regulation of glucose utilization, 116, *117*
  renal system, 123–5
  respiration and gaseous exchange, 118, *119*, 125
  role of insulin in glucose utilization, 116, *117*
  sleep, 132
  somatomedins, 118
  source of glucose, 115, 116
  surfactant, *129*, 133
  swallowing, 123, 124
  thyrotoxicosis, 128
  ultrasound image in late pregnancy, *124*
freemartin, 19, 76, *77*

galago, with central implantation and discoid placenta, 62, *63*
gastrulation, 7
gene action in embryonic development, 13, *14*, *15*, 16, 144, 145, 156, 157
genet, form of placenta, *51*
genitalia, external and internal differentiation of male and female, *95*, *96*
  see also sex determination and differentiation
gestation
  delays in, 34
  times of stages, and total, *22*
giraffe, form of placenta, *51*, **52**
glycogen, increase in embryonic cytoplasm, 11
goat
  blood oestrogen and progesterone levels before parturition, *137*
  ectoderm cells in embryo, 5, *6*
  source of progesterone, and presence of placental 17α-hydroxylase in late pregnancy, **138**
gonad, differentiation and effect of H-Y antigen, *75*
guinea pig

blood oestrogen and progesterone levels before parturition, *137*
form of placenta, *51*
interstitial implantation, 29
oestrogen apparently not necessary for implantation, 39
source of progesterone in late pregnancy, and absence of placental 17α-hydroxylase, **138**
times of stages of gestation, and total time, **22**

haemophagous organ in zonary placenta, *51*
haemophilia, 157
hamster
  blastocyst forms implantation chamber and implants interstitially, 30
  culture of fetus *in vitro*, 18
  difficulties of culturing embryo *in vitro*, 17
  oestrogen apparently not necessary for implantation, 39
hare
  form of placenta, *51*, **52** 62
  hybrid embryo with rabbit develops to blastocyst, 13, *14*
  'hatching' of blastocyst, 2, 27, 29
hermaphrodite, true, 103
hexose monophosphate oxidation pathway in embryo, 10, 11
histocompatibility antigen (H, H–Y, H–W)
  in pregnancy, 45
  in sex differentiation, *75*, 76, 78, 82, 83
horse
  blood oestrogen and progesterone levels before parturition, *137*
  development of fetal membranes, *53*, *54*
  embryo attachment on unspecialized endometrial surface, 30

horse (*cont.*)
    endometrial cup, 39, 52,
        54, *55*, 56
    placental type, *51*, **52**
    production of equine
        chorionic
        gonadotrophin, 39, 52,
        56
    time of implantation, 23
    times of stages of
        gestation, and total
        time, **22**
    hyaena, spotted,
        Crocuta crocuta, male
        and female external
        genitalia, *105*, 106
H-Y antigen, 45, *75*, 76, 78,
    82, 83
17α-hydroxylase in
    placenta, 137, **138**, 139
hyrax, form of placenta, 62

immunological relations
    between mother and
    fetus, *44*, 45, 56
impala, internal migration
    of embryos, 24
implantation
    changes in endometrium
        during, 31, 32
    control of, 34–45
    ectopic, 42, 45
    general, 27–45
    times of, **22**, 23
    types of normal, 29–31, *29*
'Infant Hercules'
    syndrome, 102
insectivore
    discoid placenta, 62
    preimplantation changes
        in embryo, 2
Isoodon obesulus, sex
    chromosomal status of
    somatic and germ cells
    differ, 85, 86

kangaroo
    characteristics of
        embryonic diapause,
        28
    diffuse placenta, 52
    embryo attaches to
        unspecialized
        endometrial surface,
        30
    paternal X chromosome
        preferentially
        inactivated, 79

kangaroo, red, *Macropus
    rufus*
    nutritional effect on
        diapause, 36, 37
    pattern of reproduction,
        36, *37*
Klinefelter's syndrome, 88,
    100

lemming, with interstitial
    implantation, 30
lemming, Arctic varying,
    Dicrostonyx torquatus,
    sex-determining
    mechanism, 82
lemming, wood, *Myopus
    schisticolor*,
    sex-determining
    mechanism, *81*, 82
lemur
    form of placenta, *51*, **52**,
        62
    type of implantation, 63

macaque, form of placenta,
    51, **52**
Macropus agilis, pattern of
    reproduction, *36*
Macropus eugenii, *see*
    tammar wallaby
Macropus robustus, effect
    of nutrition on
    embryonic diapause, 37
Macropus rufus, *see* red
    kangaroo
man
    adrenogenital syndrome
        (congenital adrenal
        hyperplasia), 102
    blood-gas exchange
        across placenta, *119*,
        120
    blood oestrogen and
        progesterone levels
        before parturition, *137*
    chromosome anomalies
        in spontaneous
        abortions, **145**
    correlation of
        spontaneous abortions
        with age, *144*
    culture of embryos *in
        vitro*, 17
    man (*cont.*)
    decidual cell reaction,
        32–4
    descent of testis, 96, *97*
    differentiation of external

and internal genitalia,
    95, *96*
distribution of glycogen
    and brown fat in fetus,
    134, 135
Down's syndrome
    (mongolism), *144*, 148,
    149
effects of too little or too
    much insulin during
    fetal life, 116, *117*
form of placenta, *51*, **52**,
    62
hydatidiform mole, *151*,
    153, *154*
implantation and
    formation of fetal
    membranes, *63*, 64,
    65, *66*, *67*, 68
implantation is
    interstitial, 30
Leonardo da Vinci
    drawing of fetus and
    placenta, *59*, 60
monosomy, *145*, 148,
    149, *150*
mosaicism in abortion,
    145
multiple births as side
    effect of
    ovulation-induction
    treatment, 18
preferential inactivation
    of defective X
    chromosome, 79
preimplantation embryo,
    2, 3, *21*
5α-reductase deficiency,
    102
tetraploidy: in abortion,
    145; in embryo, 9
times of stages of
    gestation, and total
    time, 22
triploidy in abortion,
    145, 150, *151*, 152; in
    embryo, 7, 9
trisomy 21: and maternal
    age, *144*; chromosome
    distribution, *143*
trophoblast cells invade
    endometrium, 32
twins, 18, **19**, *67*
ultrasound image of
    fetus, *124*
45, X fetus, *150*
XY female, 82
XYY male, 89

manatee, form of placenta,
52, 62
marmoset
dizygotic twins, 19, 76,
77, 78
placental anastomosis,
77, 78
marsupial
blastocyst: expands
before implantation,
29; lacks inner cell
mass, *2*, *3*, *26*
duration of embryonic
diapause, 27
placental type differs
from eutherian, 57–60
preimplantation changes
in embryo, *2*
selective inactivation of
paternal X
chromosome, 16
stage of blastocyst
formation, 3
translocation of X
chromosome on to
autosome, 78
see also bandicoot;
kangaroo; possum;
quokka; wallaby
masculinization by
administered
androgens, 103
maternal immune response
to fetus, *44*, 45, 56
medial preoptic nucleus,
sex dimorphism, *93*,
94
Meles meles, see European
badger
6-mercaptopurine, effects
of, 157
migration of embryo,
external, 24; internal
(transuterine), *23*
Miniopterus schreibersii,
implantation of
diapausal blastocyst
after hibernation, 37
mink
activation of blastocyst,
34, 35
form of placenta, *51*, 62
times of stages of
gestation, and total
time, *22*
mole, hydatidiform, *152*,
153, *154*
mole, placenta

diffuse, 52
contradeciduate, 59
mongoose, with Y
chromosome
translocated on to
autosome, 78
mongolism, *144*, 148, *149*
see also Down's
syndrome; trisomy
monkey
blood oestrogen and
progesterone levels
before parturition, *137*
capuchin, type of
implantation, *63*
form of placenta, *52*, 62
see also marmoset;
rhesus monkey
monosomy
in man, **145**, 148, 149, *150*
in mouse, 14
mosaic embryo,
experimentally
produced, 7
mosaicism in human
abortion, **145**
mouse
aneuploidy in embryo, 14
arrangement of fetal
membranes, *64*
blastocyst forms
implantation chamber,
30
cAMP levels in uterus
after stimulation of
decidual cell response,
33
chimaera: of XX and
XY embryos, 85, 86;
production of, 159–61,
173, *174*, 175, *176*, 177
cleavage rate, 2
culture *in vitro*: of
embryos, 16–18; of
fetuses, 18
development from less
than normal number
of blastomeres, 9
diapause, 27, **28**, 29, 37,
38, 41
DNA polymerase levels
in diapausing embryos,
38
ectopic implantation, 42
energy sources for
embryonic
development, **11**
endocrine control of

embryonic diapause
and implantation, 37,
38, 41
form of placenta, *51*, **52**,
62
genes active in 2- to
4-cell embryo, 13
implantation:
antimesometrial, 24;
eccentric, 30; ectopic,
42
internal migration of
embryos, 24
knowledge of cell
lineages from
transplantation
experiments, *160*, 161,
162, *163*, **164**, *165*,
166, 167, *168*, 169
monosomy, 14
parthenogenetic
development, 7
preferential inactivation
of paternal X
chromosome in
placental tissue, 79
RNA in early embryo,
11, 12
sex dimorphism in
kidney and
submaxillary gland,
103, 104
stage of blastocyst
formation, 3, 4
stimulation of decidual
cell response, *33*
teratoma, *172*, 173, 174
time of implantation, 22,
23
times of stages of
gestation, and total
time, *22*
'tobacco; *Mus
poschiavinus*, hybrids
with house mouse, 14, 15
transplantation of nuclei
in eggs, 179, *180*
triploidy in development,
7
XO chromosomal state,
87
XXY chromosomal
state, 89
mucoid (mucoprotein) coat
on marsupial embryo,
2, *26*, *27*, *28*
mule
endometrial cups in, 56

mule (*cont.*)
  preferential inactivation
    of X chromosome
    from donkey father, 79
Müllerian duct, persistent,
  syndrome, 97
Müllerian-inhibitory
  factor, 76, 91, 97, 101
Mus poschiavinus, hybrid
  embryos with house
  mouse, *14*
muscular dystrophy, 157
mustelid
  characteristics of
    embryonic diapause,
    28
  form of placenta, *51*, **52**,
    62
Myopus schisticolor, *see*
  lemming, wood

non-disjunction
  as source of human
    chromosomal
    anomalies, 147, 148
  in spermatogenesis of
    hybrid mouse, 14
nucleus transplantation in
  analysis of
  development, 178, *179*,
  180
nutrition, effect on
  embryonic diapause, 37

parthenogenetic activation,
  experimental, 7
parturition, control of
  135–41, *137, 138*, **138**,
  139
phosphoglucokinase, coded
  by X chromosome, 16
pig
  asynchrony between
    embryo and uterine
    stage, some tolerance
    shown, 42
  blood oestrogen and
    progesterone levels
    before parturition, *137*
  2-cell and 8-cell eggs, *4*
  culture of embryo *in
    vitro*, 17
  development of fetal
    membranes, *47*, 52
  elongation of blastocyst,
    2, 3, 4
  embryo: attachment on
    unspecialized

endometrial surface,
  30; transfer, *23, 24*
form of placenta, *51*, **52**
internal migration of
  embryos, 24
luteal phase oestrogen
  mainly from embryo
  trophoblast, 39
placental
  17α-hydroxylase, 138
'purple protein' in
  allantoic fluid, 43
source of progesterone in
  late pregnancy, 138
times of stages of
  gestation, and total
  time, *22*
pinealectomy, did not end
  diapause in skunk, 35
pineal gland
  denervation affects
    melatonin level, 37
  may be mediator of
    photoperiodic changes
    in corpus luteum and
    blastocyst, 35, 37
pine martin, control of
  implantation, 34, 35
Pipistrellus pipistrellus,
  extension of
  pregnancy, 34
placenta
  anastomosis in twin
    pregnancy, 19, *77*
  chorioallantoic, 46
  choriovitelline, 46, 47
  formation of fetal
    membranes, 46–68
  growth, 120
  how cortisol promotes
    oestrogen synthesis,
    138
  structure and function,
    46–68
  types of, *50–66*, **52**;
    cotyledonary, *51*,
    58–61, *60*; diffuse,
    52–9, *55*; discoid, *51*,
    61, 62–8, *64, 65, 66,
    67*; zonary, *51*
placental anastomosis, 18,
  19, 76, *77*, 78
placentome, structure and
  function, 59–61
platypus
  nutrients in egg derive
    from uterus, 46
  structure of embryo and

membranes at two
  stages of incubation,
  49
Polled (*P*), 76, 87
polyembryony, *20*, 21
  see also twins, triplets
  (etc.)
porpoise, form of placenta,
  52
possum, with diffuse
  placenta, 52
pregnancy duration, *see*
  gestation
pregnancy loss
  causes, 142–56;
    chromosomal
    abnormalities,
    143–5; *143, 144*,
    145; environmental
    agents, 145, 146;
    maternal limitations,
    145; single-gene
    defects, 145, 156, 157
  periods: early and mid
    pregnancy, 147–56,
    *148, 149, 150, 151,
    152*, **153**, *154*; late
    pregnancy, 155, 156;
    very early pregnancy,
    146, 147
pregnant mare's serum
  gonadotrophin
  (PMSG), *see* equine
  chorionic
  gonadotrophin
primate, source of
  progesterone in late
  pregnancy and absence
  of placental 17α-
  hydroxylase, **138**
prostaglandin
  in parturition, 135–40,
    137, 139
  may initiate decidual cell
    response, 34
  pathway for synthesis of
    main members, *137*
protein synthesis by
  embryo, *10, 12, 13*
puberty, 98, 99
'purple protein', in
  allantoic fluid of pig,
  43

quadruplets, 18, *20*, 21
quinacrine, for Y
  chromosome in human
  cells, 80

quintuplets, 18
quokka
　ovaries essential for
　　termination of
　　diapause, 41
　possible effect of
　　nutritional stress on
　　diapausing blastocysts,
　　37

rabbit
　aneuploidy in embryo,
　　14
　arrangement of fetal
　　membranes, *64*
　blood oestrogen and
　　progesterone levels
　　before parturition, *137*
　biochemistry of
　　blastocyst fluid, 16
　biopsy of blastocyst for
　　sex determination, 169,
　　170, 171
　blastocyst expands
　　before implantation,
　　29
　culture of embryo *in
　　vitro*, 17
　development from less
　　than normal number
　　of blastomeres, 9
　form of placenta, *51*, **52**,
　　62
　grafted testis stimulates
　　Wolffian duct and
　　inhibits Müllerian
　　duct, 90, *91*
　hybrid embryo (with
　　hare) develops to
　　blastocyst, 13, 14
　mucoprotein coat on
　　embryo, 27, *28*
　oestrogen apparently not
　　necessary for
　　implantation, 39
　preimplantation changes
　　in embryo, *2*
　RNA in early embryo,
　　11, 12
　source of progesterone in
　　late pregnancy, and
　　absence of placental
　　17α-hydroxylase, **138**
　stage of blastocyst
　　formation, 3
　times of stages of
　　gestation, and total
　　time, *22*

uterus-specific protein
　secreted in
　peri-implantation
　period, 42, 43
racoon, form of placenta,
　51
rat
　blastocyst forms
　　implantation chamber
　　and implants
　　eccentrically, 30
　blood oestrogen and
　　progesterone levels
　　before parturition, *137*
　chromosomal
　　aberrations in embryos
　　after X-irradiation or
　　treatment with
　　antimetabolites, 14
　culture *in vitro* of
　　embryo, 17; of fetus,
　　18
　embryonic diapause:
　　characteristics of, **28**;
　　duration of, 27, 29
　form of placenta, *51*, **52**
　lactational delay of
　　implantation increased
　　by reduced nutrition,
　　37
　oestrogen terminates
　　embryonic diapause
　　and initiates
　　implantation, 38
　preferential inactivation
　　of paternal X
　　chromosome in
　　placental tissue, 79
　RNA in early embryo,
　　11, 12
　sex dimorphism in
　　medial preoptic
　　nucleus, *93*, *94*
　source of progesterone in
　　late pregnancy, and
　　absence of placental
　　17α-hydroxylase, **138**
　time of implantation, 23
　times of stages of
　　gestation, and total
　　time, *22*
5α-reductase deficiency, 102
rhesus monkey
　embryo adheres to
　　endometrium, and
　　trophoblast cells
　　invade, *32*
　oestrogen apparently not

necessary for
　implantation, 39
times of stages of
　gestation, and total
　time, *22*
　type of implantation, *63*
rhinoceros, placental type,
　52
ribonucleic acid (RNA)
　embryonic messenger
　　takes over control of
　　early development, 14,
　　16
　maternal messenger
　　active in early
　　cleavage, 13, 14, 16
　synthesis in early
　　embryo, *10*, 11, 12
Rubella syndrome, 157

seal
　characteristics of
　　embryonic diapause,
　　28
　form of placenta, *51*, **52**,
　　62
Sertoli cell, secretes H-Y
　antigen, 76
sex chromatin, *80*, *81*
sex chromosome
　in snakes, *73*, *74*
　see also X and Y
　　chromosomes
sex determination and
　differentiation, 70–113
　behavioural sex, 106–11,
　　108
　brain sex, 92, *93*, *94*
　genetic sex, 72–83, *73*,
　　75, *77*, *80*, *81*
　germ cell sex, 84–90, *85*,
　　86, *88*
　gonadal sex, 83, 84
　hormonal sex, *90*, 91, *92*
　legal sex, 111, 112
　phenotypic sex, 95–104,
　　95, *96*, *97*, *98*, *100*
　reversal, in birds, 84
sex dimorphism, in kidney
　and submaxillary
　salivary gland, 103,
　104
sex ratio, influenced by
　temperature in reptiles,
　74
sex reversal gene (*Sxr*), 76,
　87, *88*
sextuplets, 18

sexual selection, 104, *105*, 106

sheep
blood oestrogen and progesterone levels before parturition, *137*
culture of embryo *in vitro*, 17
development of fetal membranes, *60*
elongation of blastocyst, *2, 3*
form of placenta, *51*, **52**
how fetal lamb controls onset of labour, 138, *139*
internal (transuterine) migration of embryos, 24
masculinization of fetus by administered androgens, 103
oestrogen apparently not necessary for implantation, 39
production of identical twins, *8, 9*, 10
some asynchrony between embryo and uterus tolerable, 42
source of progesterone in late pregnancy, and presence of placental 17α-hydroxylase, **138**
times of stages of gestation, and total time, 22

shell membrane, on marsupial embryo, 2, 26, 27, *28*

shrew, form of placenta, *51*, **52**

shrew, elephant, preimplantation changes in embryo, 2

skunk
Eastern spotted, no embryonic diapause, 35
Western spotted, control of implantation, 34, 35, *41*, 42; diapause not ended by pinealectomy, 35

snake, sex chromosomes, *73*, 74

spermatozoon chromosomally

unbalanced in hybrid mouse, *14*
X- and Y-bearing, 72, 73

*t*¹², gene functional in 2- to 4-cell mouse embryo, 13

tapir, placental type, **52**

Tarsius, type of implantation, *63*

Tay–Sachs disease, 157

tenrec, preimplantation changes in embryo, *2*

testicular feminization syndrome, 97, *100*, 101, 102

testis
descent of, 96, *97*
hormonal activities in fetus, 91, *92*
Y-linked gene determines size in mouse, 78

tetraploidy, 9, **145**, 150

thalidomide, effects of, 157

translocation in human abortuses, **145**

transplantation antigens, maternal production in pregnancy, 44, 45

tricarboxylic acid cycle in embryo metabolism, 11

triplets, 18

triploidy, 7, **145**, 150, *151*, 152

trisomy, *14, 15, 143, 144*, **145**, 147, *148, 149*, 156

trophectoderm (trophoblast), *2*, 3, 4
cells phagocytic, 33
contributes to formation of chorion, 46, *47*
gives rise to giant cells in culture, 18

Turner's syndrome, 89, *100*, 101

twins
fraternal (two-egg, dizygotic), 18, **19**, 21
identical (one-egg, monozygotic), *8, 9*, 10, 18, **19**, 21, *67*

uteroglobin, 17, 42, *43*

uterus
endometrium sensitized by oestrogen, 39, *40*
responses for

implantation, 42, *43*, 44
suitability as site for embryonic growth, *44*, 45
weight changes after stimulation of decidual cell reaction, *33*

vole, creeping, *Microtus oregoni*, no X chromosome in male germ line, 79, 84, 85

vole, mole, *Ellobius lutescens*, lacks Y chromosome, 78

wallaby
characteristics of embryonic diapause, **28**
development of fetus and membranes, 57, 58, 59
embryo attaches to unspecialized endometrial surface, 30
oestrogen apparently not necessary for implantation, 39

wallaby, agile, *Macropus agilis*, pattern of reproduction, *36*

wallaby, tammar, *Macropus eugenii*
blastocyst in diapause, 26
development of fetal membranes, 53, 57, 58
inter-related changes in embryonic diapause, *41*, 42
ovaries essential for termination of diapause, 41
pattern of reproduction, *36*
reactivation of diapaused blastocysts, 35, *36*, 37

weasel, long-tailed, control of implantation, 34, 35

whale, form of placenta, *51*, **52**

Wolffian duct, hypertrophy after graft of testis tissue, *90*, 91

X chromosome
activity in early embryo, 15, 16

X chromosome (*cont.*)
   enzymes coded on, 13, 16
   inactivation, 16, 79, *80*, 177, 178
X and Y chromosomes, in sex determination, 72, *73*, *75*, 76, 78–80, *81*, 82–4, *85*, 86–90
X-inactivation, 16, 79, *80*, 177, 178
X-irradiation, of gametes produces embryos with chromosomal aberrations, 14
XO chromosomal state, 87, 89, *100*, 101
XXY chromosomal state

in man (Klinefelter's syndrome), 88
in various animals, 80, 87
XY primary gonadal dysgenesis, *100*, 101
XYY chromosomal state, 89

Y chromosome, fluorescent in human cells, *80*
yellow, gene functional in 2- to 4-cell mouse embryo, 13

Z and W chromosomes, in sex determination, *73*, 74, 84

zona pellucida
   dissolution by protease or low pH, 29
   function to hold cleaving embryo together, 23
   loss from preimplantation embryo, *2*, 27, 29
   prevents attachment of embryo to oviduct epithelium, 27
   protects embryo from leucocytic attack, 27
   thin in marsupial eggs, 27
   use in experimental twin production, *8*, 9